MOBILITIES AND INEQUALITY

This page has been left blank intentionally

Mobilities and Inequality

Edited by

TIMO OHNMACHT
Lucerne University of Applied Sciences and Arts, Switzerland

HANJA MAKSIM
École Polytechnique Fédérale de Lausanne, Switzerland

MANFRED MAX BERGMAN
University of Basle, Switzerland

LONDON AND NEW YORK

First published 2009 by Ashgate Publishing

2 Park Square, Milton Park, Abingdon, Oxon OX14 4RN
711 Third Avenue, New York, NY 10017, USA

Routledge is an imprint of the Taylor & Francis Group, an informa business

First issued in paperback 2016

Copyright © 2009 Timo Ohnmacht, Hanja Maksim and Manfred Max Bergman

Timo Ohnmacht, Hanja Maksim and Manfred Max Bergman have asserted their right under the Copyright, Designs and Patents Act, 1988, to be identified as the editors of this work.

All rights reserved. No part of this book may be reprinted or reproduced or utilised in any form or by any electronic, mechanical, or other means, now known or hereafter invented, including photocopying and recording, or in any information storage or retrieval system, without permission in writing from the publishers.

Notice:
Product or corporate names may be trademarks or registered trademarks, and are used only for identification and explanation without intent to infringe.

British Library Cataloguing in Publication Data
Mobilities and inequality. - (Transport and society)
  1. Transportation - Social aspects 2. Equality 3. Social
  mobility
  I. Ohnmacht, Timo II. Maksim, Hanja III. Bergman, Manfred
  Max
  303.4'832

Library of Congress Cataloging-in-Publication Data
Ohnmacht, Timo, 1979-
  Mobilities and inequality / by Timo Ohnmacht, Hanja Maksim, and Manfred Max Bergman.
       p. cm. -- (Transport and society)
  Includes bibliographical references and index.
  ISBN 978-0-7546-7495-5
  1. Social mobility. 2. Social structure. 3. Equality. I. Maksim, Hanja. II. Bergman, Manfred Max. III. Title.

  HT612.O46 2009
  305.5'13--dc22

2008045423

ISBN 978-0-7546-7495-5 (hbk)
ISBN 978-1-138-25433-6 (pbk)

Contents

List of Figures and Tables	*vii*
Notes on Contributors	*ix*
Foreword	*xiii*

Mobilities and Inequality – An Introduction   1
Timo Ohnmacht, Hanja Maksim and Manfred Max Bergman

PART I: THEORY, CONCEPTS, AND FINDINGS ON MOBILITIES AND INEQUALITY

1   Mobilities and Inequality – Making Connections   7
    Timo Ohnmacht, Hanja Maksim and Manfred Max Bergman

2   Unequal Mobilities   27
    Katharina Manderscheid

3   Life Course Inequalities in the Globalisation Process   51
    Hans-Peter Blossfeld, Sandra Buchholz and Dirk Hofäcker

4   Metaphors of Mobility – Inequality on the Move   75
    Jonas Larsen and Michael Hviid Jacobsen

PART II: EMPIRICAL APPLICATIONS

5   Mobilities and Social Network Geography: Size and Spatial Dispersion – the Zurich Case Study   99
    Andreas Frei, Kay W. Axhausen and Timo Ohnmacht

6   Social Integration Faced with Commuting: More Widespread and Less Dense Support Networks   121
    Gil Viry, Vincent Kaufmann and Eric D. Widmer

7   Here, There, and In-Between: On the Interplay of Multilocal Living, Space, and Inequality   145
    Nicola Hilti

8	Class Divides within Transnationalisation – The German Population and its Cross-Border Practices *Steffen Mau and Jan Mewes*	165
9	Residential Location, Mobility and the City: Mediating and Reproducing Social Inequity *Markus Hesse and Joachim Scheiner*	187
10	Mobility and the Promotion of Public Transport in Johannesburg *Ursula Scheidegger*	207

Index *219*

List of Figures and Tables

Figures

3.1	Globalisation and rising uncertainties in modern societies	53
5.1	Distribution of the number of relationships	109
5.2	Example social geography measured by confidence ellipse	110
5.3	Distribution of the social network geometries measured as 95% confidence ellipses (km^2)	112
6.1	Illustration of binding and bridging social capital	124
6.2	Illustration (to scale) of the network spatial expansion according to a weak (2 km) or strong (50 km) commuting distance of the respondent	135
6.3	Illustration (to scale) of the network spatial expansion according to the education level and residential context of the respondent	139
8.1	Educational attainment and transnational mobility	172
8.2	World maps by education: above – high educational level; below – low/average educational level	173
8.3	Average distance of transnational short- and long-term mobility (by educational level)	175
9.1	Location of the study areas in the region of Cologne	191

Tables

3.1	Life courses in the globalisation process	61
5.1	Socio-demographic comparison between the characteristics of the Zurich respondents and the Zurich population	107
5.2	Number of social relationships by socio-demographic characteristics	111
5.3	Parameter estimates for the negative binominal regression of the number of contacts	112
5.4	Parameter estimates for the Tobit regression of the logarithm of the size of the 95% confidence ellipses and the associated Probit model of the Cragg approach	115
6.1	Summary of the used scale variables	127
6.2	Frequency (in %) of the respondents' socio-demographic variables	128

6.3	Regression analysis of network expansion on different variables related to the respondent (unstandardised regression coefficients)	129
6.4	Regression analysis of the activation of the emotional support ties (in ‰) Ego-Alters on different variables related to the respondent (unstandardised regression coefficients)	132
6.5	Regression analyses of the activation of the emotional support ties (in ‰) Alter-Alter on different variables related to the respondent (unstandardised regression coefficients)	134
8.1	Determinants of involvement into transnational social practices (OLS regressions)	178
9.1	Life style dimensions and items	193
9.2	Location preference dimensions and items	194
9.3	Structure of analysis: types of models	195
9.4	Models of residential location choice – 'basic models' and 'extended models' vs 'state models' and 'process models'	199

Notes on Contributors

Kay W. Axhausen is Full Professor of Transport Planning at the Institute for Transport Planning and Systems (IVT) of the Swiss Federal Institute of Technology (ETHZ) in Zurich, Switzerland. His research fields include spatial accessibilities with regard to dynamics of commuting and land use in Germany and Switzerland 1970 to 2000 as well as travel impacts of social networks and networking tools.

Manfred Max Bergman is a Full Professor of Sociology at the University of Basle, Switzerland at the Department of Sociology. His research interests include mixing qualitative and quantitative methods and the issue of mobilities and modernity.

Hans-Peter Blossfeld holds the Chair of Sociology I at Bamberg University and is the Director of the State Institute for Family Research at Bamberg University. He works on social inequality, youth, family, and educational sociology, labour market research, demography, social stratification and mobility, the modern methods of quantitative social research and statistical methods for longitudinal data analysis. Currently, he is interested in the flexibilisation of work in modern societies, the division of domestic work in the family, the development of individual competences, and the formation of educational decisions in early school careers.

Sandra Buchholz is a research scientist at the University of Bamberg, Chair of Sociology I. She holds a PhD in Sociology from Bamberg University (Dr. rer. pol.). She is interested in labour market research, international comparison, life course research, industrial relations, and quantitative methods.

Andreas Frei has studied civil engineering at the Swiss Federal Institute of Technology Zurich. He is presently research officer and doctoral student at the Institute for Transport Planning and Systems (IVT) of the Swiss Federal Institute of Technology (ETHZ). His main research focus is in modelling of transport behavior and social networks.

Markus Hesse is Professor of Urban Studies at the University of Luxembourg, Geography and Spatial Planning Research Centre. His main research fields include principles of urban and regional development; European urban development and policy; flows and mobilities in an urban context.

Nicola Hilti is a research associate and PhD candidate at the ETH Wohnforum – Centre for Research on Architecture, Society and the Built Environment (ETH CASE) at the ETH Zurich. She holds a master's degree in sociology and communication studies from the University of Vienna. Currently, her major research focus is multilocality as a social and cultural practice of everyday life.

Dirk Hofäcker is a research scientist at the University of Bamberg, Chair of Sociology I and at the Institute for Family Research at the University of Bamberg (ifb). His interests are in family sociology, comparative labour market research, life course research, attitudinal research, and quantitative methods in the social sciences.

Michael Hviid Jacobsen holds a PhD in sociology. He is a senior lecturer in sociology and director of studies at the Department of Sociology, Social Work and Organisation at Aalborg University, Denmark. He has a research focus on transformation of modernity.

Vincent Kaufmann is Professor of Urban Sociology and Mobility at Ecole Polytechnique Fédérale de Lausanne (EPFL). His fields of research are mobility and urban life styles, links between social and spatial mobility and public policies of land planning and transportation.

Jonas Larsen holds a PhD in human geography. He was previously a lecturer in sociology at the Department of Sociology, Social Work and Organisation at Aalborg University, Denmark, but now lectures at Roskilde University in the Department of Environment, Society and Spatial Change. He is interested in mobilities and tourism studies.

Hanja Maksim is a sociologist, research assistant and doctoral student at LaSUR Laboratoire de Sociologie Urbaine in Lausanne, Switzerland. Her research focus is on urban planning policy, mobility potential and social inequalities.

Katharina Manderscheid holds a PhD in sociology from the University of Freiburg i.Br. Currently she works at the Centre for Mobilities Research at Lancaster University as a visiting research fellow, holds a scholarship of the Swiss National Foundation and she is a member of the Cosmobilities Network. Her main research fields include mobility, space, topology, spatial planning, sustainability, social inequality, social stratification, social justice and gender.

Steffen Mau is Professor of Sociology (Political Sociology and Comparative Social Research) at the University of Bremen. Currently, he acts as Dean of the Bremen International Graduate School of Social Sciences. He works in the fields of comparative welfare research, social inequality and European integration.

Jan Mewes is research assistant at the Institute of Sociology at the University of Bremen, where he works in the project 'Transnationalization of Social Relations'. His PhD project focuses on the general relation between social inequality and the structure of personal communities.

Timo Ohnmacht is a sociologist, research officer at the Institute of Tourism at the Lucerne University of Applied Sciences and Arts and a doctoral student at the University of Basle, Department of Sociology, Switzerland. His research focus is on the interrelation of mobilities and modernity as well as leisure and tourism travel.

Ursula Scheidegger holds a MA in Political Studies, is a political scientist and a part-time lecturer and researcher at the Department of Political Studies, University of the Witwatersrand, Johannesburg, South Africa. Her research interests are urban development and governance.

Joachim Scheiner holds a PhD in geography. He is a senior researcher and lecturer at the Faculty of Spatial Planning, Department of Transport Planning at the Technische Universität Dortmund. His main research interests centre around travel behaviour in the context of social and spatial structures and processes.

Gil Viry is a sociologist and a research assistant in sociology and doctoral student at LaSUR (Laboratory of Urban Sociology) at Ecole Polytechnique Fédérale de Lausanne (EPFL), Switzerland. He is interested in the interactions between job-related spatial mobility and the private sphere. His research focus is on links between social networks and spatial mobilities.

Eric D. Widmer is Professor at the Department of Sociology of the University of Geneva (Switzerland), with an appointment at the Lemanic Center for Life Course and Lifestyle Studies (Pavie). His long-term interests include family relations, life course research and social networks.

This page has been left blank intentionally

Foreword

The Cosmobilities Network considers itself as a sort of incubator for interdisciplinary and international mobility research. It is a precious realm for open discussions and new ideas and insights into the very modern nature of mobility and its relevance for the modernisation of the world.

Many of the ideas and thoughts collected in this compilation here were firstly presented during the 4th Cosmobilities Conference in Basle, Switzerland, in September 2007. Entitled 'Mobilities, Space and Inequality', this conference gathered people from very different disciplinary backgrounds, concerned with very different issues. The book in your hands shows how the editors and the organisers of the conference perfectly handled the rising complexity and transdisciplinarity that shape contemporary mobility research. International experts came together for lively discussions on social, physical, cultural, and virtual mobilities and their interdependencies and interrelations on social inequality, and space. The essences from these exciting debates are now available in this book. The authors contribute to the too often neglected dimension of social inequality in mobility research. This is the achievement of this collection – to rearrange these topics for the agenda of a cosmopolitan mobility research.

This book deserves a huge number of open-minded readers. And hopefully they will help to push forward a more specific research agenda on inequality in relation to geographical space and mobility.

Sven Kesselring
Cosmobilities Network
Technische Universität München

This page has been left blank intentionally

Mobilities and Inequality – An Introduction

Timo Ohnmacht, Hanja Maksim, and Manfred Max Bergman

Mobilities and Modernity

In the last few decades, societies have become visibly more dynamic, heterogeneous and internationalised, due to vastly increased levels and forms of mobilities, understood in this context as displacements of goods, information and people (Urry 2008). Dynamics initiated through mobilities appear not solely in advanced modern societies; current developments around the world demonstrate how the ubiquity of various mobilities is transforming the globe. To name a few: the 'automobilisation' of India's society is planned through the '$2,000 car' which is being promoted by the Indian automotive company Tata Motors; in Africa, the '$100 laptop' was initiated to equip Internet cafes across the continent; in Cape Town a new public transport system is planned to move soccer fans safely to the FIFA Soccer World Cup in 2010; in Bangladesh new financial mobility in the form of microcredits flows to the unemployed and to poor workers who were not previously considered creditworthy. These examples illustrate the significant changes that societies are experiencing due to the various forms of modern mobilities. Considering the worldwide spread of various mobilities, one can argue that mobilities are tantamount to individualisation, reflexivisation and globalisation. According to Canzler et al. (2008, 3), mobilities have to be conceived as 'a basic principle of modernity' or, metaphorically speaking, could be regarded as 'the flip side of late modernity' (Fotel 2006, 746).

Mobilities, Inequality and Academia

Dynamics caused by modernisation, globalisation, migration and social change affect the structuring process of both the social fabric and geographical space. Such dynamics are due, in no small part, to recent achievements in communication and transport technologies which offer new possibilities for social arrangements, spatial settings and functional overlapping. Various 'mobilities' have attracted the attention of scientific disciplines. A considerable amount of literature can be found in the area of social research and theory called mobility studies. This field has been established with the explicit aim of studying the mobile aspects of social life, especially with regard to the interplay between society, space and mobilities. In this research field, mobilities and patterns of inequality are conceived of as reciprocally related. Social inequality refers to the differential distribution of, and access to, scarce yet desirable goods or resources and inequality can also mean disparity of opportunity to maintain or improve status. By way of example of such disparities, richer societies have been shown by experts to possess a greater range of mobility systems, which are related to the capacity to maintain or improve status. It is such capacity that allows a company to deploy a new mobility system to reach a new core market for an old product, or a professional to relocate to Zurich following a new job offer.

Mobilities have taken centre stage in analysing and explaining dynamics in societies. Anthropology, cultural studies, geography, migration studies, science and technology studies, tourism and transport studies and sociology all deploy understandings of the dynamics connected to mobilities and inequality to complement and develop the aims of their research fields. Over recent years, several book-length studies have emerged on the issue of mobilities and inequality (Kaufmann 2008; Urry 2008; Mignot and Rosales-Montano 2006; Le Breton 2008; Hine and Mitchell 2003; Gaffron et al. 2001; Daphne 1992), in addition to edited books (Rajé et al. 2004) and special issues of academic journals (Macdonald and Grieco 2007; Kaufmann et al. 2007). The backbone of the present volume is a contribution to these examinations of contemporary dynamics within societies through a focus on various specific mobilities. In so doing, this publication presents a threefold contribution to recent debates on the issue of mobilities and inequality:

- First, the volume is explicitly focused on how mobility is linked to classical forms of social inequality. The classical conception of social inequality refers to the differential distribution of wealth, income, educational attainment and status. These resources can be tangible or intangible and they can differ according to quantity or quality. Thus, the book provides a framework to think in more detail about the interplay between mobilities and classical forms of social inequality.
- Second, the contributions highlight 'new forms' of inequality which interconnect with various mobilities. Until the 1970s, inequality studies primarily focused on vertical differentiation, such as poverty and social class. Within recent decades, the increased availability of choice in societies has lead to more complex patterns of inequality structures within society, notably including the increasing variability of lifestyles, attitudes, opinions and values. Regarding these horizontal expansions of social inequality, for instance on a micro-societal level, social inequality may vary within the same stratum of society and, therefore, have intended and unintended effects on mobilities.

- Third, the volume extends the recent discussions by applying 'mobile methods' (Urry 2008). For instance, in contemporary debates, the notions of 'motility' (Kaufmann, 2002; Kaufmann et al. 2004) and 'opportunity space' (Canzler and Knie 1998) have become prominent analytical tools to differentiate between potential and fulfilled mobility. The latter has to be regarded as movement. Against this background it will be argued that the opportunity and capacity to fulfil or to avoid various mobilities is also highly linked to 'classical' and 'new' forms of inequality.

In sum, the book is a contribution to recent debates within the fields of, inter alia, mobility studies, politics and transport policy. We hope that through investigating forms of inequality against the background of mobilities we can make a substantive contribution to a better understanding of mobility opportunities and movements of people, goods and information. Due to the pervasiveness and perseverance of difference in social position, status and power, the focus on inequality and its relation to various mobilities is a fundamentally important pursuit for social policy as it relates to unequally distributed chances to participate in – and also avoid – mobility. Hence, the contributions are relevant for practitioners and researchers in different positions dealing with the question of mobilities and inequality.

References

Canzler, W., Kaufmann, V. and Kesselring, S. (2008), 'Tracing Mobility: An Introduction', in Canzler, W., Kaufmann, V. and Kesselring, S. (eds), *Tracing Mobilities: Towards a Cosmopolitan Perspective* (Aldershot: Ashgate).
Canzler, W. and Knie, A. (1998), *Möglichkeitsräume: Grundrisse einer modernen Mobilitäts- und Verkehrspolitik* (Wien: Böhlau).
Daphne, S. (1992), *Gendered Spaces* (Chapel Hill, NC: University of North Carolina).
Fotel, T. (2006), 'Space, Power, and Mobility: Car Traffic as a Controversial Issue in Neighbourhood Regeneration', *Environment and Planning A* 38:4, 733–48.
Gaffron, P., Hine, J. and Mitchell, F. (2001), *The Role of Transport in Social Exclusion in Urban Scotland* (Edinburgh: Scottish Office Central Research Unit).
Hine, J.P. and Mitchell, F. (2003), *Transport Disadvantages and Social Exclusion: Exclusionary Mechanisms in Transport in Urban Scotland* (Ashgate: Aldershot).
Kaufmann, V. (2002), *Re-thinking Mobility: Contemporary Sociology* (Aldershot: Ashgate).

Kaufmann, V. (2008), *Les Paradoxes de la Mobilité* (Lausanne: Presses polytechniques et universitaires romandes).

Kaufmann, V., Bergman, M.M. and Joye, D. (2004), 'Motility: Mobility as Capital', *International Journal of Urban and Regional Research* 28:4, 745–65.

Kaufmann, V., Kesselring, S., Manderscheid, K. and Sager, F. (2007), 'Mobility, Space and Inequalities', *Swiss Journal of Sociology* 33:1, 5–6.

Le Breton, E. (2008), *Domicile-travail: les salariés à bout de souffle* (Paris: Les carnets de l'info).

Macdonald, K. and Grieco, M. (2007), 'Accessibility, Mobility and Connectivity', *Mobilities* 2:1, 1–14.

Mignot, D. and Rosales-Montano, S. (2006), *Vers un droit à la mobilité pour tous. Inéqualitiés, territoires et vie quotidienne* (Paris: La documentation française).

Rajé, F., Grieco, M., Hine, J. and Preston, J. (eds) (2004), *Transport, Demand Management and Social Inclusion: The Need for Ethnic Perspectives* (Aldershot: Ashgate).

Urry, J. (2008), *Mobilities* (Oxford: Blackwell).

PART I
Theory, Concepts, and Findings on Mobilities and Inequality

This page has been left blank intentionally

Chapter 1
Mobilities and Inequality – Making Connections

Timo Ohnmacht, Hanja Maksim and Manfred Max Bergman

Introduction

Few themes have been as central to the social sciences as the conceptualisation and study of inequality and the distribution of social and economic resources. Yet, while generations of researchers have explored a multitude of theories, conceptualisations and measurements of social stratification, the conjunction with mobilities[1] – in this context understood as displacements of goods, information and people – has not been explored in great detail (cf. Kaufmann et al. 2007).

Based on a recognition of ongoing processes of pluralisation, differentiation and the increasing influence of mobilities within all societies, it is crucial to turn attention to investigation of the interconnectedness between patterns of inequality and mobilities. We are interested in following such basic questions as: What types of mobilities cause patterns of inequality structures and dynamics? What patterns of inequality structures lead to certain kinds of mobilities? And what are the consequences on further mobilities?

Before exploring the interrelations between mobilities and inequality, we have to consider first, the concepts of social stratification and inequality and, second, the notion of mobilities. A rigorous consideration of these concepts will reveal potential interrelations and mutual consequences with regard to modern societal dynamics. Therefore, this chapter provides first, definitions and, second, a conceptual linkage of both key notions. Furthermore, we argue that thinking explicitly about the potential of being mobile enlarges the theoretical understanding of the mutual relation between mobilities and inequality. Thus, we apply the theoretical approach of 'motility', defined as the capacity to move, and its interrelation to inequality and fulfilled mobilities

---

1 Due to the fact that the notion of *mobilities*, which implies various forms of movement, digital information exchange, means of transportation and so forth, has been established within mobility studies we use the plural form of *mobility* (Urry 2007). For the same reason, it is also possible to write about *inequalities*. But we avoid pluralisation and apply the more common used notion of 'patterns of inequality'.

(Kaufmann et al. 2004). Finally, we provide concrete examples from current studies in transport studies with regard to social exclusion.

Making Connections

On Social Stratification

Social stratification refers to unequally distributed resources, i.e. wealth, status, prestige, or power, within a social system (Erikson and Goldthorpe 1992; Goldthorpe 1985). The most widespread social stratification approach relates to social class. Three main strands can be distinguished.

First, the idea of distinct social classes has been around since at least Ancient Rome and has had a tremendous impact on the social and political sciences since Karl Marx. Marx did not pay much explicit attention to the notion of class in his main work, but it is a latent conception within his social conflict theory. He distinguished between the proletariat and bourgeoisie as groups defined by their different relationship to the ownership of means of production throughout history (Marx 1990 [1867]).

Another important tradition relates to the subjective assessment of status or prestige, most often measured from attitude scales on occupational titles (Blau and Duncan 1967). This strand emphasises professional occupational titles as the primary defining criterion of social position. Recently, Bergman and Joye (2001) have discussed contemporary stratification research and compared various stratification schemata based on occupational titles.

Third, stratification can also be based on the links between resources and social networks. Accordingly, networks give access to socially desirable, yet scarce, resources while, concurrently, access to resources position individuals within networks (Wassermann and Faust 1994; Stewart et al. 1980; Bergman et al. 2002).

Despite the differences in defining and measuring social stratification, most of these approaches agree that social stratification is not invariant across time and place. In fact, the social fabric contains a form of mobility in itself. Dynamics which evolve within the social fabric are usually termed 'social mobility'. It is defined by Sorokin (1927, 133) as

> any transition of an individual or social object of value – anything that has been created or modified by human activity – from one social position to another.

Conceptually speaking, social mobility is divided into a vertical and a horizontal dimension. Vertical social mobility implies the forms of ascending or descending the social ladder of economic, political and occupational participation; horizontal social mobility stands for the transition or shifting of an individual or social object from one social group to another without change in its vertical level (Sorokin 1970; Morgan et al. 2006). These dynamics can be understood on a micro-societal level, first, as intragenerational social mobility within an individual's life time, e.g. the increase or decrease as well as the horizontal shift of cultural and economic capital of social actors over their lifetime. Second, they can be approached from the perspective of intergenerational mobility, e.g. the transfer of cultural, social and economic capital between generations. Third, societal mobility can also be understood in the form of social change and fluidity due to, for instance, diminishing agricultural and skilled manual sectors and a general societal change in quality of life, health and prosperity of a population over time (also known as the 'elevator or lift effect').

According to this theoretical background, our understanding of social stratification is based on the placement of individuals within 'social space'. For us, 'social space' implies at least two dimensions. Differences relating to sociodemography and socio-economy, such as gender, age, income, educational attainment form one axis, and the space of lifestyles, attitudes, opinions and values forms the other (Bergman 1998). In this ideal-typical understanding, the two axis are often considered orthogonal, but it may be stressed that these two axis are also often considered interdependent (e.g. Erikson and Goldthorpe 1992). In summary, the various interactions of these two axes – in a simplified understanding – build up status, prestige, power and so on. Finally, one can conclude that social stratification is a multi-layered concept, constructed by the mentioned dimensions. To show how social stratification is produced and reproduced by access to various resources, in the following, we expand on social inequality.

On Social Inequality

Social inequality and social stratification are interrelated; social stratification produces institutionalised patterns of inequality and patterns of inequality produce stratification structures. Within this reciprocal understanding, social stratification refers, as we have seen, to the unequal distribution of scarce yet desirable goods or resources, whereas inequality means the disparity of opportunity or capacity to maintain, or improve, status. Thus, social inequality is to be understood as institutionalised difference in access to resources. Along these lines, Stamm et al. (2003) suggest distinguishing between notions of difference which are implicit to social stratification on the one hand and inequality, which stands for the institutionalised access to goods and resources, on the other. Gender identity, for example, does not necessarily lead to a higher or lower income *per se*. Instead, income differences are often based on socially constructed and institutionalised inequality structures. Inequality implicates at least two interrelated conceptual levels: socio-economic inequality, which includes poverty, welfare and social stratification; political and cultural inequality, which relates to issues including citizenship, governance and social justice, and thus implies multicultural and collective identity concerns. In fact, one can argue that the concept of social exclusion and integration is the fusion between socio-economic and political inequality structures and dynamics.

In summary, patterns of social inequality have their origin in our understanding of various dimensions of social stratification. The economic and social segregation of the social fabric leads to different patterns of inequality (Berger and Hradil 1990; Berger 1998; Erikson and Goldthorpe 1992). We distinguish between classical and new forms of inequality. Classical inequality involves distinctions of class, status, wealth, prestige and so forth which are mediated by gender, income and education. New forms of inequality follow from a greater awareness of increased global complexity and the existence of a greater range of choices for individuals; these forms are therefore defined by such things as consumption, lifestyle, identity dynamics and so forth. This horizontal diversification has lead to an increase in heterogeneity within social stratification studies (e.g. variations in attitudes, opinions and values and behaviour in the same status level) (Gross 1994; Schulze 2005). Conceptually speaking, social differentiation produces and reproduces dynamic forms of inequality due to both concrete placement within social space as well as vertical and horizontal social mobility.

On Mobilities

In this section we focus on various mobilities in terms of the displacement of people, goods and information. We begin by discussing how the interplay between mobilities and societies emerged on the research agenda within the social sciences in recent decades.

Mobilities on the research agenda of the social sciences Based on the debates in the field of mobility studies, it appears that 'all the world is on the move' (Sheller and Urry 2006, 207), such as asylum seekers, soldiers, business people, backpackers, commuters, musicians, prostitutes, refugees, the early retired, young mobile professionals, terrorists, tourists, international students, members of diasporic cultures, financial transfers, information on the Internet, international non-governmental organisations (INGOs) and evermore intergovermental organisations at the regional and global level and so forth. Despite such a picture, however, a more nuanced presentation is needed to clarify who or what is moving, or staying put, how they are doing so and how often, for what reasons, and under what circumstances and influences.

To disaggregate the often stated holistic view of ubiquitous 'mobilities', authors like Beck (2008), Urry (2007), Sheller and Urry (2006) and Kaufmann et al. (2004) proclaim a new 'mobilities paradigm' in the social science in the sense of Kuhn's understanding of 'paradigm' (Kuhn 1996). Due to this paradigm shift, they argue that it is necessary to develop 'mobile methods' to understand the recent mobility dynamics in greater detail. Thus, new questions have emerged on the research agenda and new methodological tools have become available to explore the dynamics between mobilities and society. As a consequence, large-scale movements of people, objects, capital and information across the world, as well as more local processes of daily transportation, movement through public space, mobility and the travel of material things within everyday life are now taking centre stage in the social sciences. Researchers have developed theoretical and methodological approaches to explore difficulties, opportunities and discoveries with regard to mobilities. There are approaches which focus explicitly on manifest mobilities on the one hand, while others explore the potentials to fulfill or even avoid mobilities on the other. The former deals with concrete movements of peoples, information and goods. It has to be understood in this context as 'manifest mobilities'. The latter considers the ability of individuals and groups to take advantage of displacement and also widens the perspective to include the capacity to be immobile. Therefore this approach expands the dynamic of mobility potential to both application and refusal. It has to be understood in this context as 'the transformation from latent to manifest mobilities'. In the following, we deploy these two forms of mobilities and then mirror them against the background of inequality.

Narratives on 'Manifest Mobilities' Urry (2008) identifies various forms of mobility under the notion of 'mobilities' (see also Urry 2000). First, physical travel of people (e.g. work, leisure, family life, pleasure, migration and escape). Second, physical movement of objects (e.g. flows of economic goods, objects delivered to producers, consumers and retailers). Third, imaginative travel (e.g. through images of places and peoples via television and other media, 'armchair travelling'). Fourth, virtual travel (e.g. often in real time on the Internet, city tours). Fifth, communicative travel (e.g. through person-to-person messages via email, letters, telephone, fax and mobile phone). The implementation of these diverse and interacting mobilities has many consequences for social actors, and they are linked to the placement of social actors within social space. Moreover, Larsen et al. (2006) argue that these intermittent mobilities form and reform social life.

Narratives on the transformation from 'Latent Mobilities' to 'Manifest Mobilities' To fully conceptualise various forms of mobilities, one has to be aware what is and what could be mobile. In the field of mobility studies, an approach has developed to conceptualise opportunity space for mobility based on the theoretical notion of *motility* (Kaufmann 2002). By the term motility, Kaufmann et al. (2004, 750) refer to

> the capacity of entities (e.g. goods, information, or persons) to be mobile in social and geographic space, or as the way in which entities access and

appropriate the capacity for socio-spatial mobility according to their circumstances.

On a micro-societal level, motility refers to the capacity to relocate both socially and geographically. It has to be understood as 'how an individual or a group take possession of the realm of possibilities for mobility and build on it to develop personal projects' (Flamm and Kaufmann 2006, 168). Motility consists of three main layers, which form the capacity of movement (Kaufmann et al. 2004, 750):

- *Access* refers to the range of possible mobilities according to place, time and other contextual constraints, and may be influenced by networks and dynamics within territories. Access is constrained by options and conditions. The options refer to the entire range of means of transportation and communication available, and the entire range of services and equipment accessible at a given time. The conditions refer to the accessibility of the options in terms of location-specific cost, logistics and other constraints. Obviously, access depends on the spatial distribution of the population and infrastructure (e.g. towns and cities provide a different range of choices of goods and services), sedimentation of spatial policies (e.g. transportation and accessibility) and socio-economic position (e.g. purchasing power, position in a hierarchy or social network).
- *Competence* includes skills and abilities that may directly or indirectly relate to access and appropriation. Three aspects are central to the competence component of motility: physical ability, e.g. the ability to transfer an entity from one place to another within given constraints; acquired skills relating to rules and regulations of movement, e.g. licences, permits, specific knowledge of the terrain or codes; and organisational skills, e.g. planning and synchronising activities including the acquisition of information, abilities and skills. Competence is multifaceted and interdependent with access and appropriation.
- *Appropriation* refers to how agents (including individuals, groups, networks, or institutions) interpret and act upon perceived or real access and skills. Appropriation is shaped by needs, plans, aspirations and understandings of agents, and it relates to strategies, motives, values and habits. Appropriation describes how agents consider, deem appropriate and select specific options. It is also the means by which skills and decisions are evaluated.

The interplay between these three elements forms the foundation of mobilities. Hence it ought to describe the transformation from manifest to latent mobilities described above. Motility is a concept to distinguish between the opportunity and the implementation of social and geographical mobility. On a micro-level, the extent to which mobility opportunities can be realised depends on resources of an actor as well as her or his intention of implementation. This is also the case for groups, networks, institutions, or goods. In summary, mobilities are the materialisation of motility.

Relations between Mobilities and Inequality

In the following section we discuss the interplay between inequality and mobilities, based on the following questions: how are mobilities structured by inequality, how do mobilities structure inequality and how are they reciprocally interconnected and what are the consequences?

How are Mobilities Structured by Inequality?

The transformation from latent to manifest mobilities is highly related to inequality structures. The multiplicity of various social strata gives rise to distinctive mobilities, mediated by forms of inequality in terms of monetary resources and time. Urry (2007, 67) argues that these mobilities vary highly according to status, wealth, prestige, power and so forth.

> [A]ccess is unequally distributed but the structuring of this inequality depends *inter alia* on the economics of production and consumptions of the objects relevant to mobility, the nature of civil society (the association and organisations beyond the economy state), the geographical distribution of people and activities, and the particular mobility-systems in play and their forms of interdependence. (Urry 2008, 17)

From this it follows that there are financial, physical, organisational, attitude-driven, spatial and temporal components that lead to mobilities. These components are profoundly marked by distinctions of gender, age, ethnicity and so forth. Individuals' mobilities are determined by their ability to move due to cognitive, technical, societal and economic capacities. There is no doubt that the unequal equipment with 'mobility capital' leads to different opportunities to fulfill both everyday life-spaces and long-term mobility. Classical forms of inequality, including demographic and economic distributions, lead to differential access to, and exclusion from, various mobility systems. Accessibility to opportunities, services and social networks is therefore profoundly marked by distinctions of class, gender, age and ethnicity mediated through various mobilities (Urry 2007, 67).

How do Mobilities Structure Inequality?

Structural changes within societies have led to various demands and even requirements of being mobile. Examples include the demands on a household of two working people who work in different areas (Holmes 2004; Walby 1997) to organise complex daily activity programmes (for example, work, hobbies and to accompany children), the issue of access to the employment market in a context of unemployment growth and the increase in fixed term contracts and the implications of the increasing requirement to travel as part of one's job. Thus, mobilities can lead to patterns of inequality as well as to the narrowing of difference between groups in the inequality structure. Manderscheid and Bergman (2008), for instance, examine links between social differentiations and spatial demands of Swiss citizens. While the distribution of professional opportunities is unequally distributed in Switzerland, the Swiss public transportation system provides the means of access to the relevant localities. This infrastructure helps various people living in peripheral regions to bridge spatial distance and improve their chances of converting their social and cultural capital into economic gain. Beside the question of commuting, mobilities form and reform social life more generally (Urry 2007, 47) since social interaction depends on the ability to be mobile in terms of the usage of means of transport and communication. This is particularly relevant in a society of increasingly spatially dispersed social networks (Ohnmacht et al. 2008). The ability to stay in touch with significant others, such as important friends, family and so forth, for example, enables social actors to maintain their social capital. On a micro-societal level, attitudes, opinions and values towards mobility tools (i.e. cars, trains, planes, bicycles and so forth) may also contribute to or be linked to social stratification. A preference for distinctive cars, bikes, planes and so forth has to be seen in conjunction with horizontal stratification. According to Hall (1999, 183), inequality may be manifested in the mode of transport. A comfortable and air-conditioned coach ride in Central Asia, for example, may reflect income and life-style differences between the traveller and the indigenous culture.

How are Mobilities and Inequality Reciprocally Interconnected and What are the Consequences?

The reciprocal relationship between mobilities and inequality is best illustrated through the notion of motility. By focusing on the transformation from latent to manifest mobilities, we can understand how access, skills and knowledge that go along with the willingness or even unwillingness to take advantage of mobilities may be transformed into advantage and power. According to Kaufmann et al. (2004, 754), motility and 'its conceptualization as a form of capital which can be mobilised and transformed into other types of capital (i.e. economic, human and social capital) allows motility to make original contributions in the research area relating to social inequality and social change'. Along these lines, as Manderscheid and Bergman (2008, 7) put it:

'concept of motility seems useful in illuminating the link between mobility and social inequality, as motility can be used to explain different social and spatial trajectories in relation to one's spatial and social position'.

Moreover, Flamm and Kaufmann (2006, 168) reason that:

sociological analysis today can no longer function without an in-depth analysis of the role of mobility in social integration and its implication in terms of social differentiation, or even exclusion.

Against this background of interpretation, we can ascertain that different degrees of motility can lead to a certain 'spatialisation' in terms of the underlying mobility potentials. Kaufmann et al. (2004) reason that motility is a form of capital itself. Therefore, it can be understood to be conceptualised on the same level as economic, social, or cultural capital (Bourdieu 1984). Importantly, motility may be transformed into, and be gained from the transformation of, other forms of capital. Motility seems to be a useful theoretical approach in understanding, for example, that the potential for geographical mobility is an important factor for gaining social and economic access in situations such as in relocations due to an attractive job offer. Patterns of inequality go hand in hand with access, competence and appropriation of mobilities on one hand (Bacqué and Fol 2007), while, in certain contexts, it can lead to the freedom to resist mobilities on the other.

In summary, the most adequate way to analyse societies under the conditions of globalisation and global complexity is to employ and develop 'mobile methods' in order to understand how these processes relate to the five forms of mobilities (Urry 2007) and motility (Kaufmann et al. 2004), as well as how these mobilities interact with each others. This is because, in our understanding, movement and rigidity within the social fabric is a strong factor in shaping contemporary mobilities. Reciprocally, mobilities produce and reproduce the social fabric as well as inequality feedbacks with the social fabric. As a consequence, every social formation has its specific 'social spatialisation' (see Shields 1991). The discussed two 'mobile methods' have to be linked to different forms of inequality and on the same time have to be seen as a pattern of inequality. Mobilities, both latent and manifest, are unequally distributed throughout society, while inequality engenders and reinforces mobilities. Consequently, mobilities must be understood as both a result of, and a contributing factor to, social stratification and social inequality.

The Example of Transport Studies: Social Exclusion and Access

So far we have focused on inequality and mobilities from a conceptual perspective. In this section, we draw upon a specific research field to demonstrate the interplay between mobilities and inequality. We assess the range of inequality patterns in terms of mobilities observed in the field of transport studies where disadvantages in relation to transport systems have been examined.

Theoretical Debates

Definitions of transport-based social exclusion and accessibility According to Lyons (2003, 342), the issue of social exclusion has been a significant but relatively minor part of the current research agenda within transportation studies. Social exclusion can be seen as one consequence of inequality. Overall, social exclusion is a multilayered concept which has led to numerous definitions within transport studies, reflecting the various ways in which the two can be understood to relate to each other (for a overview, see Rajé 2003, 322).

One such definition, proposed by Kenyon et al. (2006a, 210) defines mobility-related exclusion as

> [t]he process by which people are prevented from participating in the economic, political and social life of the community because of reduced accessibility to opportunities, services and social networks, due in whole or in part to insufficient mobility in a society and an environment built around the assumption of high mobility.

In this understanding, social exclusion is considered as a lack of accessibility to fulfil social participation and to attain goods and services. In general, the notion of accessibility is understood in transportation science as a destination-based measure which indicates both the place's connectedness to means of transport and the place's attractiveness in terms of opportunities that can be realised there (Ben-Akiva and Lerman 1985). As Hillman et al. (1973) put it this is the 'get-at-ability of a destination'. But accessibility has not only to be understood as a place-based measure, but rather as a measure of the 'get-to-ability of a social actor', as, for instance, discussed by the notion of motility. Such considerations are dealt with in the context of social exclusion and transport in the form of approaches that include an actor perspective of accessible opportunities which is denoted as 'social accessibility' (e.g. Handy and Niemeier 1997; Götz 2007). Thus, transport may influence social exclusion processes due to high degrees of accessibility to opportunity structures, such as libraries, schools, hospitals, shopping malls, gyms and so forth as wells as contact with significant others, such as friends, spouse, kinship and so forth.

Detection and measures of transport-based social exclusion and accessibility To detect the threat of social exclusion due to a lack of transport systems various measures have been discussed in recent debates. According to Grieco (2006), there is still a need for adequate measurable proxies for parameters of exclusion related to transport concerns. To date, the following approaches have been suggested for detection and measures of transport-based social exclusion and accessibility.

Church et al. (2000) reason that social exclusion can be detected in seven main categories: first, physical exclusion such as hearing and visually impaired people (Matthews and Vujakovic 1995), mobility-disabled people (Kitchin 1998), the elderly and so forth. Second, geographical exclusion, which means inaccessibility due to peripherality and therewith linked poor provision of public transport services. Third, exclusion from facilities, for instance due to the absence of school buses to the next secondary schools in rural area. Fourth, economic exclusion, such as low income groups who can not afford to maintain their spatially distant contacts. Fifth, time-based exclusion, for instance the lack of time to overcome distance by travel may be a problem for double career couples working and living in different cities. Sixth, fear-based exclusion, for instance the refusal to use public transport due to the fear of the presence of other people. Finally, space exclusion, for example inability to obtain a visa to travel in a specific country. These dimensions are clearly interrelated.

To analyse social inclusion or exclusion, Grieco (2006) proposes three main dimensions: First, place-based measures, such as opportunities and services within the immediate surrounding of a person. Second, social-category based measures, such as social stratification within a community which give indications as to social needs (e.g. percentage of childbearing women who are provided with antenatal facilities). Third, person-based measures, such as the individual public transport user's profile of journey needs. Similarly, Preston and Rajé (2007) use a matrix of area accessibility, area mobility and individual mobility to identify concentrated and scattered manifestations of social exclusion and inclusion.

Church et al. (2000, 197) criticise the fact that official statistics which measure social exclusion include a number of indicators, including unemployment, income support, health, education, crime in the neighbourhood, overcrowding and lack of basic amenities, but, apart from non-car ownership, none that inquire about the respondents' travel behaviour. Although car ownership is used as a proxy for poverty, its absence does not always imply that a household is poor, at least in dense urban areas with extensive public transport infrastructures and scarce parking.

Empirical Findings

To date, research in the domain of transport studies has shown various effects of inequality on travel behaviour which indicates social exclusion. To illustrate we focus on empirical findings for the cases of Germany, Switzerland and the United Kingdom.

Inequality and travel The 2002 transport survey 'Mobility in Germany' reveals a strong positive relationship between traffic volume and social indicators, such as income, employment status, quality of residential area and level of education. In addition, there is evidence that women travel less, but have far more complex trip chains than men do due to obligations arising from family, work and shopping (Kunert et al. 2004, 124).

These findings about the connection between income and travel are in line with those of the 'Swiss Microcensus on Travel Behaviour 2005', which shows that the higher the income, the more often people are on vacation (ARE/BFS 2007). Furthermore, for Switzerland is known that the highest income group travels twice as fast than the lowest income group in terms of the interrelation of income and mean velocity of the daily kilometres travelled (ARE/BFS 2001). Possible causes for this striking difference between income groups may be, first, due to differences in the choice of destination (e.g. leisure activities in the immediate surrounding of the residential location versus nation-wide leisure trips). Second, differences may occur due to the mode of transport being determined by income (for instance low-income groups may make greater use of human powered mobility while higher-income groups make more frequent usage of cars and planes).

For the United Kingdom, the research report 'Making the Connection: Transport and Social Exclusion', produced by the British government's Social Exclusion Unit, reveals similar results in terms of income as a mobility modifier (Social Exclusion Unit 2002). In accordance with Cass et al. (2005, 543), it concludes, among other things, that

> young people with a driving licence are twice as likely to get jobs as those without; that nearly one-half of 16–18 year olds experience difficulty in paying for transport to get to their place of study; that almost one-third of car-less households have difficulty in accessing their local hospital; that children from the lowest social class are five times more likely to die in road accidents than those from the highest social class; and that twice as many people without a car find it hard to see their friends in comparison to those with access to car.

Social exclusion, inclusion and travel A subject of early discussion within transport studies was how accessibility to employment and urban services can indicate the quality of urban living. As Wachs and Kumagai (1973, 438) point out, social exclusion and inclusion due to

> physical mobility are major contributors to social and economic inequality, and that differences are systematically related to such variables as age, race, sex, and location of residence within the region.

By applying an accessibility analysis to Los Angeles, Wachs and Kumagai (1973, 455) concluded that:

> Even the simplified analysis which was performed demonstrated that significant differences exist in accessibility to employment and to health care opportunities, and that these differences seem to be related to socio-economic status and to spatial location of communities within the region.

McCray and Brais (2007) have researched the role that transport plays in fostering social exclusion for low-income women. Their qualitative findings reveal important insights into the effect of transportation upon social exclusion. They show, for instance, that a female interviewee living in a suburban area was unable to join a library despite desiring to do so due to an inability to afford the cost of the bus to get there. Moreover, with regard to perceived inequality, the respondents stated that transport planning favours the improvement of transportation infrastructure within areas populated by higher-income groups (McCray and Brais 2007, 404).

Rajé (2003) looked at the relation between congestion charging and social inclusion. Based on research of the perception of people likely to be affected by road pricing they conclude that lower income groups, which are less likely to be committed to car-based travel, profit from increasing funding of bus services or improved public transport cross-financed by congestion charging.

In her work, Fotel (2006) highlights the contribution of automobility to processes of inequality. She argues that the impact from the negative side-effects of automobility, whether congestion, pollution, or noise, are unequally distributed throughout society. For instance, while there are neighbourhoods who have the power to limit traffic, others do not (Fotel 2006, 734). In fact, in Scandinavia, the lowest social group – measured by labour-market position – is more exposed to noise from traffic than higher social groups (Fotel 2006, 735). In summary, according to Fotel (2006, 735), urban patterns of inequality have a very real and direct effect in mobility materialisation upon the social sphere.

Policy measures It has been shown that transport has the potential to act as a gatekeeper. Transportation companies have developed various measures, such as the provision of low steps into buses, trams and so forth for public transport users.

On a policy level, policy makers have to be aware that people who are excluded from 'mainstream society' can suffer from exclusion from transport systems and, therefore, exclusion from society. The evidence on the relationship between transport and social inclusion and exclusion contains useful information for policy measures. In particular such measures need to address the fact that, following the pattern of the awareness that famine is not caused by lack of food but by lack of access to food, mobilities-related social exclusion is not caused by a lack of mobilities but by lack of access to it.

In terms of monetary cost and travel, policy agendas may focus on disadvantaged areas in order to help the 'transport-poor' by providing, for instance, comprehensive public transport services that may support both place-based and social accessibility. One example of measures that follow from an awareness of, for example, gender difference is the provision of free child transport for mothers in public transport who are often accompanied by their children when they do their shopping (Grieco 2006, 18). It is possible that the reduction of transport costs enhances social inclusion (Preston and Rajé (2007, 153). Thus, increased transport budgets for welfare recipients could help to maintain mobility. Furthermore, for socially dispersed social networks it is desirable to increase encounters mediated by information and communication technologies and the increase of proximate contacts (Kenyon et al. 2006a, 2006b).

In general, policy measures may help to provide increase facilities for services and goods. Furthermore, urban sprawl has an impact on social exclusion. For Britain, it is shown that population density has fallen from 1000 people per hectare to just over 50 in 100 years, although the population itself has doubled (Power 2001, 735). It is concluded that current densities are not able to sustain buses, shops, banks or schools within walking distance. Thus, the use of automobiles has became essential. According to Power (2001, 741), nothing would contribute more to protect people from transport-related social exclusion due to a lack of monetary means than greater urban density.

Summary and Conclusion

In this chapter, we have deliberately painted with a broad brush. Our first aim was to introduce the notions of mobilities and inequality. Second, we discussed some recent issues using the example of the field of transport studies and social exclusion. It has been shown that inequality underpins contemporary forms of mobilities and, likewise, that mobilities are a contributing factor to patterns of inequality. By using the notion of motility we discussed how mobilities should also be considered as a form of capital, alongside economic and cultural capital. Thus, mobilities are interchangeable with economic and cultural access. People's spatial access is, for example, increasingly important for social inclusion. In general, social inequality includes dimensions which are significantly a matter of overcoming constraints of space and time by various mobilities to gain access to work, families, friendship, (leisure) services and goods (Cass et al. 2005). Therefore, socio-spatial inclusion and exclusion relates to inequality in the availability and usage of travel and communications.

We conclude that being mobile and immobile are important factors of social differentiation and a generator of patterns of inequality. A close look at mobilities and inequality is a way of framing the spatialisation of cultural, political and social relations. Against the background of the discussion of recent findings in the research field of mobilities and inequality, one might pose the questions: *why is a further look at the relation between inequality and mobilities necessary?*

By referring to the empirical results of travel behaviour research discussed above, such as that the poorer the person, the less distance they tend to travel, we can make a simple first answer which is that our current understanding does not elucidate which social processes produce and reproduce these distinctive forms of inequality; the shown key figures do not provide a sufficient explanation of the ways in which mobilities are a central contributors to stratification in contemporary modern societies.

Furthermore, though we do know that, on a macro level, the richer the society, the greater the range of mobility systems (Urry 2008, 16), this finding is only valuable on an aggregate level. Thus, it remains unclear to what extent and under what circumstances micro-societal mobilities are directly shaped by social dynamics and directly linked to inequality in terms of the access to transport systems.

Against this background, we argue that the link between mobilities and inequality has not been sufficiently explored in social theory and empirical research; much remains to be explored in this field. In concluding this section, we want to draw attention to possible further research steps. In the first place, new data needs to be collected and analysed to help separate the causes from the effects. In so doing one has to keep methodological challenges in mind. For instance, it is well known that the marginalised are more reluctant to participate in quantitative surveys (for travel diary exercise see Richardson et al. 1995). Second, improved modelling methods need to be developed to grasp the underlying dynamics for a better understanding and for more efficient policy measures. Third, new theories need to be cultivated for a better insight into underlying effects that patterns of inequality have on mobilities. Fourth, a cross-disciplinary dialogue on the mix of concepts, methods and evidence about the connection between mobilities and inequality may be a first step to take. We hope that recent steps will provide a starting point in such a dialogue and that that the findings and discussions presented in the chapter will stimulate ideas for further research agendas.

References

ARE/BFS (2001), *Mobilität in der Schweiz, Ergebnisse des Mikrozensus 2000 zum Verkehrsverhalten* (Bern and Neuenburg: Federal Office for Spatial Development and Swiss Federal Statistical Office).

ARE/BFS (2007), *Mobilität in der Schweiz, Ergebnisse des Mikrozensus 2005 zum Verkehrsverhalten* (Bern and Neuenburg: Federal Office for Spatial Development and Swiss Federal Statistical Office).

Bacqué, M.-H. and Fol, S. (2007), 'L'inégalité face à la mobilité: du constat à l'injonction', *Swiss Journal of Sociology* 33:1, 89–104.

Beck, U. (2008), 'Mobility and the Cosmopolitan Perspective', in Canzler, W., Kaufmann, V. and Kesselring, S. (eds), *Tracing Mobilities: Towards a Cosmopolitan Perspective* (Aldershot: Ashgate), 25–36.

Ben-Akiva, M.E. and Lerman, S.R. (1985), 'Disaggregate Travel and Mobility Choice Models and Measures of Accessibility', in Hensher, D.A. and Stopher, P.R. (eds), *Behavioural Travel Modeling* (London: Croom Helm), 654–79.

Berger, P.A. (1998), *Alte Ungleichheiten, neue Spaltungen* (Opladen: Leske und Budrich).

Berger, P.A. and Hradil, S. (eds) (1990), *Lebenslage – Lebensläufe – Lebensstile (Soziale Welt – Sonderband 7)* (Göttingen: Verlag Otto Schwatz).

Bergman, M.M. (1998), 'A Theoretical Note on the Differences between Attitudes, Opinions, and Values', *Swiss Political Science Review* 2:4, 81–93.

Bergman, M.M. and Joye, D. (2001), 'Comparing Social Stratifications Schemas: Camsis, CSP-CH, Goldthorpe, ISCO-88, Treiman, and Wright', *Cambridge Studies in Social Research* 9, 1–37.

Bergman, M.M., Lambert, P., Prandy, K. and Joye, D. (2002), 'Theorization, Construction, and Validation of a Social Stratification Scale: Cambridge Social Interaction and Stratification Scale (CAMSIS) for Switzerland', *Swiss Journal of Sociology* 28:1, 7–25.

Blau, P.M. and Duncan, O.D. (eds) (1967), *The American Occupational Structure* (New York: Wiley).

Bourdieu, P. (1984), *Distinction: A Social Critique of the Judgment of Taste* (Cambridge, MA: Harvard University Press).

Cass, N., Shove, E. and Urry, J. (2005), 'Social Exclusion, Mobility and Access', *Sociological Review* 53:3, 539–55.

Church, A., Frost, M., and Sullivan, K. (2000), 'Transport and Social Exclusion in London', *Transport Policy* 7:3, 195–205.

Erikson, R. and Goldthorpe, J.H. (1992), *The Constant Flux. A Study of Class Mobility in Industrial Societies* (Oxford: Clarendon).

Flamm, M. and Kaufmann, V. (2006), 'Operationalising the Concept of Motility: A Qualitative Study', *Mobilities* 1:2, 167–89.

Fotel, T. (2006), 'Space, Power, and Mobility: Car Traffic as a Controversial Issue in Neighbourhood Regeneration', *Environment and Planning A* 38:4, 733–48.

Goldthorpe, J.H. (1985), *Die Analyse sozialer Ungleichheit: Kontinuität, Erneuerung, Innovation* (Opladen: Westdeutscher Verlag).

Grieco, M. (2006), 'Accessibility, Mobility and Connectivity: The Changing Frontiers of Everyday Routine', *European Spatial Research and Policy* (Special Issue) 78:6, 1360–80.

Gross, P. (1994), *Die Multioptionsgesellschaft* (Frankfurt/Main: Suhrkamp).

Götz, K. (2007), *Freizeit-Mobilität im Alltag oder Disponible Zeit, Auszeit, Eigenzeit – warum wir in der Freizeit raus müssen* (Berlin: Duncker und Humblot).

Hall, D.R. (1999), 'Conceptualising Tourism Transport: Inequality and Externality Issues', *Journal of Transport Geography* 7:3, 181–8.

Handy, S.L. and Niemeier, D.A. (1997), 'Measuring Accessibility: An Exploration of Issue and Alternatives', *Environment and Planning A* 29:7, 1175–94.

Hillman, M., Henderson P. and Whalley, P.J. (1973), *Transport Realities and Planning Policy* (London: PEP).

Holmes, M. (2004), 'An Equal Distance? Individualisation, Gender and Intimacy in Distance Relationships', *Sociological Review* 52:2, 180–200.

Kaufmann, V. (2002), *Re-Thinking Mobility: Contemporary Sociology* (Aldershot: Ashgate).

Kaufmann, V. (2008), *Les Paradoxes de la Mobilité* (Lausanne: Presses polytechniques et universitaires romandes).

Kaufmann, V., Bergman, M.M. and Joye, D. (2004), 'Motility: Mobility as Capital', *International Journal of Urban and Regional Research* 28:4, 745–65.

Kaufmann, V., Kesselring, S., Manderscheid, K. and Sager, F. (2007), 'Mobility, Space and Social Inequality', *Swiss Journal of Sociology* 33:1, 5–6.

Kenyon, S., Lyons, G. and Rafferty, J. (2006a), 'Transport and Social Exclusion: Investigating the Possibility of Promoting Inclusion through Virtual Mobility', *Journal of Transport Geography* 10:3, 207–19.

Kenyon, S., Lyons, G. and Rafferty, J. (2006b), 'Social Exclusion and Transport in the UK: A Role for Virtual Accessibility in the Alleviation of Mobility-Related Social Exclusion?', *Journal of Social Policy* 32:3, 317–38.

Kitchin, R. (1998), "Out of Place', 'Knowing One's Place': Space, Power and the Exclusion of Disabled People', *Disability and Society* 13:3, 343–56.

Kuhn, T.S. (1996), *The Structure of Scientific Revolutions* (Chicago, IL: University of Chicago Press).

Kunert, U. et al. (2004), '*Mobilität in Deutschland: Ergebnisbericht*' (home page), <http://www.mobilitaet-in-deutschland.de/03_kontiv2002/publikationen.htm>, accessed 12 March 2008.

Larsen, J., Urry, J. and Axhausen, K.W. (2006), *Mobilities, Networks, Geographies* (Aldershot: Ashgate).

Lyons, G. (2003), 'The Introduction of Social Exclusion into the Field of Travel Behaviour', *Transport Policy* 10:4, 339–42.

Manderscheid, K. and Bergman, M. (2008), 'Spatial Patterns and Social Inequality in Switzerland – Modern or Postmodern?', in Pflieger, G.,

Pattaroni, L., Jemelin, C. and Kaufmann, V. (eds), *The Social Fabric of the Networked City* (London: Routledge), 41–65.

Marx, K. (1990 [1867]), *Capital: Critique of Political Economy* (London: Penguin Classics).

Matthews, M.H. and Vujakovic, P. (1995), 'Private Worlds and Public Places: Mapping the Environmental Values of Wheelchair Users', *Environment and Planning A* 27:7, 1069–83.

McCray, T. and Brais, N. (2007), 'Exploring the Role of Transportation in Fostering Social Exclusion: The Use of GIS to support Qualitative Data', *Networks and Spatial Economics* 7:2, 397–412.

Morgan, S.L., Fields, G. and Grusky, D. (eds) (2006), *Mobility and Inequality: Frontiers of Research in Sociology and Economics* (Stanford, CA: Stanford University Press).

Ohnmacht, T., Frei, A. and Axhausen, K.W. (2008), 'Mobilitätsbiografie und Netzwerkgeografie: Wessen soziales Netzwerk ist räumlich dispers?', *Swiss Journal for Sociology* 31:1, 131–64.

Power, A. (2001), 'Social Exclusion and Urban Sprawl: Is the Rescue of Cities Possible?', *Regional Studies* 8:35, 731–42.

Preston, J. and Rajé, F. (2007), 'Accessibility, Mobility and Transport-related Social Exclusion', *Journal of Transport Geography* 15:3, 151–60.

Rajé, F. (2003), 'The Impact of Transport on Social Exclusion Processes with Specific Emphasis on Road User Charging', *Transport Policy* 10:4, 321–38.

Richardson, A.J., Ampt, E.S. and Meyburg, A. (eds) (1995), *Survey Methods for Transport Planning* (Melbourne: Eucalyptus Press).

Schulze, G. (2005), *Die Erlebnisgesellschaft: Kultursoziologie der Gegenwart* (Frankfurt/Main: Campus).

Sheller, M. and Urry, J. (2006), 'The New Mobilities Paradigm', *Environment and Planning A* 38:2, 207–26.

Shields, R. (1991), *Places on the Margin. Alternative Geographies of Modernity* (London: Routledge).

Social Exclusion Unit (2002), 'Making the Connections: Transport and Social Exclusion: Interim Findings from the Social Exclusion Unit (SEU)', http://archive.cabinetoffice.gov.uk/seu/downloaddoc4f62.html, accessed 12 March 2008.

Sorokin, P. (1927), *Social Mobility* (New York: Harper and Brothers).

Sorokin, P. (1970), *Social and Cultural Dynamics* (Boston: Porter Sargent Publishers).

Stamm, H., Lamprecht, M. and Nef, R. (2003), *Soziale (Un)gleichheit in der Schweiz: Strukturen und Wahrnehmungen* (Zürich: Seismo).

Stewart, A., Prandy, K. and Blackburn, R. (1980), *Social Stratification and Occupations* (London: Macmillan).

Turner, J. and Grieco M. (2000), 'The Impact of Transport Investments Projects upon the Inner City: A Literature Review', *Time and Society* 9:1, 129–36.

Urry, J. (2000), *Sociology beyond Societies: Mobilities for the Twenty-First Century* (New York: Routledge).
Urry, J. (2007), *Mobilities* (Oxford: Blackwell).
Urry, J. (2008), 'Moving on the Mobility Turn', in Canzler, W., Kaufmann, V. and Kesselring, S. (eds), *Tracing Mobilities: Towards a Cosmopolitan Perspective* (Aldershot: Ashgate), 13–24.
Wachs, M. and Kumagai, G.T. (1973), 'Physical Accessibility as a Social Indicator', *Socio-Economic Planning Science* 7:5, 437–56.
Walby, S. (1997), *Gender Transformation* (London: Routledge).
Wassermann, S. and Faust, K. (1994), *Social Network Analysis: Methods and Applications* (New York: Cambridge University Press).

This page has been left blank intentionally

Chapter 2
Unequal Mobilities
Katharina Manderscheid

Introduction

Mobilities and social inequalities are complexly interwoven. Neither mobility nor social inequality are set, static or given categories but both concepts are infused with meaning, power and contested understanding, linked with specific practices and social and spatial arrangements, thus continuously reiterated and reproduced (cf. Cresswell and Uteng 2008, 1). Following the mobilities paradigm (Sheller and Urry 2006, 212), *mobilities* are understood as embracing physical movement of people and objects as well as movement of images and information.[1] In order to understand these mobilities and their links with stratification and power, it is important to include the ability to move and the impact of potential mobilities (Kaufmann 2002; Sager 2006). Furthermore, movement is not meaningful in itself but gains social significance through discursive constitutions and links with symbolic profit (Cresswell 2006; Frello 2008). These social embeddings of mobilities correlate and interact with corresponding material productions of landscapes – produced by spatial planning (Healey 2004, 64; Richardson and Jensen 2003), transportation policies as well as other material practices. Thus, technologies of transportation and communication do not lead in a one dimensional way to a specific usage but their accessibility and appropriation is embedded in the emergence of practices which forms specific socio-spatial arrangements (Peters 2006; Shove 2003; Wajcman 2008).

In large parts of sociology, *social inequalities* are understood and operationalised as unequal distributions of valuable resources, amongst which income and professional position especially have gained the most attention of social researchers (cf. Tienda and Grusky 1994). Little emphasis is put on the question of why some resources are thought to be socially significant and others not, how these unequal distributions are produced and reproduced and what

---

[1] In the literature, the terms *mobility* and *movement* are used largely interchangeably. Still, mobility refers slightly more to the 'social processes around movement' (Drewes Nielsen 2005, 53) thus denoting 'the socially produced motion' (Cresswell 2006, 3) whereas movement focuses more on the 'act of displacement ... between locations' (Cresswell 2006, 2f). The use of these terms in this paper follows this terminology without emphasising or elaborating on their slightly different accentuation.

impacts they have on people's everyday lives and chances. Less mainstreamed relational and multidimensional approaches to social inequalities emphasise their multifaceted nature beyond the economic or professional sphere. Furthermore, inequalities appear to be contingent, connected to specific spatial and historic contexts; they are continuously produced and reproduced through power relations within the different fields of the social and enacted through stratified attitudes, practices and lifestyles (Bourdieu 2000; Bradley 1996; Fraser 2003).

Geographic location, connectedness and the ability to move seem to have gained increased significance within these social power struggles over significant differences since spatial relations are undergoing large transformations characterised as 'time-space compression' (Harvey 1990, 204ff), the emergence of the 'networked space of flows' (Castells 2002) and as the end of the 'infrastructural ideal' resulting in 'splintering urbanism' (Graham and Marvin 2001, 90ff). Yet, despite the broad public, political and academic recognition of a growing transnational interconnectedness of social relations, apart from conceptual papers and few qualitative studies (Beck 2007; Nowicka 2006; Weiss 2005), the national space remains the self-evident research frame of studies on social inequalities.[2] Furthermore, most theories and analyses on social inequality still largely ignore mobilities and space as dimensions and forces within the process of reproducing inequalities, which seems partly linked with the focus on the distributional aspect rather than the everyday practical enactment of social inequality (cf. Jiron 2007). This research gap in inequality studies corresponds with a fundamental lack of systematic theorisation and empirical analysis of the interrelations and complex interactions of mobilities and inequalities. Moreover, although within the body of literature on mobilities and globalisation it is prominently stated that mobilities are a crucial force within the processes of social structuring and stratifying (Bauman 2000, 120; Castells 2002; Urry 2007, 185ff), a lot of research in this vein is still waiting to be carried out.

In the following conceptual contribution I will outline four levels at which the interweaving of social inequalities, power relations, mobilities and space can be studied. These four levels represent somewhat steps from a basic description of unequal movement to a more complex and elaborated understanding of movement, their technological material foundation, evolving culturally embedded meaning and social practices in their stratifying and stratified dimensions. Starting with a short sketch of the current restructurings of the spatial infrastructures of connectedness, which form the base of movement in geographic space, I look at the unequal impacts of these material structures. The focus on actual and observable movement is the most basic understanding of mobilities and dominates transportation studies. In a second step I will

---

2 The concept of nationally and territorially framed societies has been criticised also by the spatial and the mobilities turn (cf. Massey 2005; Urry 2000).

look for the social foundations of unequal abilities of movement. Concepts like motility and network capital seem suitable to capture the fundamental interlinkings of social and spatial structures, schemes of perception and valuation and patterns of movement. Moreover, the ability as well as the potential to move are outlined as fundamental mechanisms within the reproduction of unequal chances. Thirdly, I point to the social constitution of mobilities as being always infused with social meaning. This highlights the contingent and relational character of movement, which makes sense only in a broader frame of socially produced significances and meanings. Finally, I will look more closely at the link between technologies of transportation and communication and social practices. Whereas the emergence of these technologies is seen largely as the cause of the acceleration and increased connectedness of the social, more elaborated approaches highlight the co-emergence and co-constitution of technologies, meanings and practices. The focus on these links and interactions of the stratifying and excluding moments of material infrastructures, practices and meanings of passages (Peters 2006) provides a sophisticated approach to studying unequal mobilities.

By discussing these different perspectives on mobilities and inequalities I show that a broad conceptualisation of these two terms and their study from different angles is necessary to capture their complex interactions. For this purpose, the following may serve as conceptual framework for further research. Moreover, a corresponding understanding of social inequalities as inherent in and produced by mobilities allows for a more adequate understanding of the shifting socio-spatial frames which appear to be no longer congruent with territorial nation states, yet have not dissolved into universal borderlessness (cf. Walby 2001).

Unequal Spaces and Movement

The first and most basic perspective regarding inequalities focuses on the material constitution of space as the networked foundation of mobilities and on the observable unequal movement of people. Drawing on the spatial turn in social sciences, rather than equalised with the surface of the earth, *space* is understood as socially made through movement, thus forming and synthesising relations (Löw 2001; Massey 2005): 'As people, capital, and things move they form and reform space itself' (Sheller and Urry 2006, 216). Further, since people, capital and things do not move on the same tracks in the same rhythm, they form different spatialities which appear ever less congruent with predefined social territories.

Globalisation is often understood as the era of immediacy and instantaneousness, when there is no longer a need to go anywhere because everywhere is connected to everywhere (Bauman 2000, 118). But a closer look reveals that not all places are equally well connected, as the material infrastructures underlying movement are not equally distributed geographically. And, as has been pointed out by transportation studies, different social groups do not move equally smoothly through geographic space (e.g. Hine and Mitchell 2003; ARE 2006).

A considerable body of approaches focuses on the shaping and changing of the underlying material bases of mobilities, that is, the structures of connectivities. The recent emphasis in social sciences on the relational and ongoing constitution of space in opposition to its conceptualisation as separate socio-spatial units (Massey 2005, 3–7) seems to correspond also to a paradigm change in national spatial planning strategies (cf. Graham and Marvin 2001; Harvey 1990; Manderscheid and Bergman 2008). During the era of the national welfare systems, spatial planning in many European countries aimed predominantly at eradicating regional differences – in economic development or the provision of public infrastructure, – in order to provide comparable social conditions for all of the country's citizens (e.g. ROG (German Spatial Planning Law), Abschnitt 1 §1 Abs. 6). This corresponds to an understanding of society as a territorially framed social unit. More recently, spatial planning strategies seem to account for as well as contribute to increasing interconnections across and beyond national borders, thus producing what is commonly referred to as networked space (Healey 2004, 64f; Richardson 2006). These changes are rooted in an increase of transnational political governance and in fundamental restructurings of economic spatial relations (Harvey 1990, 284–307), especially with the dominance of the contemporary international financial system (e.g. Thrift 1996, 213–55).

Very prominently, Castells (2001; 2002) describes this new globalised type of spatiality as characterised by *networks* and *flows*. By flows, Castells refers to the repetitive sequences of interchange and movement between positions within different systems of the social, for example, the economic, political and symbolic spheres (Castells 2001, 467). The material base of these flows consists of a network of technological means of communication and transportation. Of particular importance within these networks and flows are the nodes which form increasingly powerful positions unhinged from their hinterland, as has been by Sassen (2001) argued for the system of global cities. Thus, the networked space, with its nodes, re-articulates and shifts the opposition of centre and periphery, which appears to be no longer a metric continuum but can be found in geographic proximity. Adjacent to or overlapping with the dominant space of flows exist various spaces of places, characterised by Castells as scattered, fragmented and unconnected. Although the lives of the majorities still take place in these rather localised spatialities, the logic of the space of flows – that is, the flows of capital, information and goods and the lifestyles of the global elite – shapes its typical hubs such as airports, hotels and conference centres as well as the materialities and cultural offerings of cities and the economic and spatial policies of regions and countries. Thus, the two spatial logics imply a steep hierarchy of power relations which continues to increase. Whereas the connectivities of the central nodes in space continue to increase, less profitable areas and social groups in between the networked nodes and hubs tend to get increasingly disconnected, bypassed by infrastructure as well as socio-cultural investment (cf. Graham 2002; Graham and Marvin 2001; Noller 1999; Ronneberger et al. 1999; Sassen 2001).

Irrespective of Castells' spatial typology of flows and places being slightly simplified, it highlights the paradigm change from national territorial spatialities to a plurality of networked spaces. Accordingly, the focus on spatial inequality and its manifestation in privileged and unprivileged positions seem to be no longer sufficiently framed by national borders or operationalised through metric distances. Rather, what appears crucial is the connectivity of or lived relationality between locations in space (cf. Günzel 2007).

Unequal material infrastructures of connectivity are socially relevant since they prescribe possible *patterns of movement* of people, goods and information. Typically, transportation studies focusing on actual movement apply rational choice models in which moving is understood as a derived demand, that is, physical movement takes place in order to access options, goods or activities at another location (Hine and Mitchell 2003, 22ff). In this view, the relational position in geographic space concerning various opportunity structures and the availability of private or public means of transportation contributes to one's social position as it allows or hinders the 'access' of jobs, healthcare and education facilities, etc. (cf. DfT 2000). Material spatial inequalities are thus linked to social inequality operationalised as access to activities, resources and goods by means of transportation (Urry 2007, 187).

Many studies try to assess the impact of spatial position on patterns of movements by applying a comparative quantitative research design. Amongst others, using a large data set for the Copenhagen Metropolitan Area, Næss (2006) for example compares the impacts of urban structures on travel behavior regarding commuting, everyday trips and leisure travel. He finds that 'living in a dense area close to central Copenhagen contributes to less travel, a lower share of car driving and more trips by bike or on foot. Conversely, living in the peripheral parts of the metropolitan area contributes to a higher amount of transport and a lower share of travel by non-motorized modes' (Næss 2006, 219). So, not only the spatial extent of trips but also their number and mode seems to be largely shaped by a household's or individual's relation with the material environment and infrastructures. City centres are typically characterised by a high concentration of infrastructures and opportunities, thus representing some kind of node which appears to reduce the need to travel. Similarly, for three urban areas in Scotland, Hine and Mitchell (2003) highlight the opportunity-reducing and thereby excluding impact stemming from lack of transportation.

Apart from the available mobility structures, it is different preferences, needs and attitudes which are deemed to determine the selection of opportunities and the corresponding differences of mobility patterns which tend to vary with, amongst other things, gender, age, social status or education (cf. Næss 2008, 174; Holden 2007, 115ff). This means that spatialities are not only shaped by material infrastructures but also by socio-demographically differentiated needs In this vein, many studies analyse gender differences in transportation. Typically, it is found that women make more trips, travel shorter distances to work and use public transport more often than they use a car or use non-motorised modes (cf. Flade and Limbourg 1999; Root et al. 2002). These differences are explained by gender roles which put women in charge of childcare and by a more local orientation of women which confines their choice of workplaces (Næss 2008, 178f; Manderscheid and Bergman 2008). This means that there is an interaction between socio-demographic and spatial variables.

In a similar way, again finding more interactions between spatial and social structures, the different mobility patterns of men and women in connection with other socio-demographic variables are analysed with regards to their purpose. Moreover, there are numerous studies of the transportation patterns of other socio-demographically defined groups. These studies describe and illustrate the multiple and different lived spatialities of different social groups. Yet the rational choice foundation of many of these studies allows only a very limited understanding of the interplay of social inequalities, spatial infrastructures and mobility patterns, since they cannot theorise the link between socio-spatial structures and individual or collective practices.

A more sociologically-informed strand of studies applies differentiated lifestyle approaches to stratification. The modes and motivations of mobilities are seen as largely shaped by these orientations and, further, in contrast to the rational choice black-box of individual preferences, lifestyles are understood as shaped by social, spatial and individual conditions and experiences and shaping social practices (Bourdieu 2000; Schulze 1992). For example, on the basis of data from the Frankfurt suburban region, Jetzkowitz et al. (2007, 164) demonstrate that everyday mobility and residential mobility correlate with various forms of living arrangements and lifestyle orientations. In this approach, residence location appears at least partly as a lifestyle expression rather than a given independent variable, thus at the same time impacting on and resulting from social differentiation and stratification. Still, social groups with different life orientations and different needs resulting from them may choose the same residential area for different reasons (cf. Manderscheid 2004; Scheiner and Kasper 2002) and may correspondingly differ concerning their mobility patterns and the lived spatialities constituted through them (Götz et al. 2003).

This first group of approaches to inequality and mobilities traces the co-existence of a multiplicity of spaces (Löw 2001; Massey 2005) and thereby highlights that the latter are not congruent with set territorial units such as urban neighbourhoods, cities, regions or countries, but are highly pre-structured by material networks of connections and enacted through actual movement. And these carried out movements differ from other social dimensions, explained by a set of needs or lifestyle orientations. Due to the unequal spatial distribution of infrastructures and opportunities, the residential location in combination with means of transportation impact on social life chances. Therefore, spatial planning and transportation can be understood as forces moderating or aggravating social inequalities by allowing for easier or more restricted access to services, activities and goods (Manderscheid and Bergman 2008). Still, by seeing the endpoints of movement in a bundle of predefined needs and locations largely independent of the means and chances to access them, these approaches fail to see their socio-cultural embedding. The changing character of these endpoints has been emphasised by Cass et al. (2005, 542):

> What is necessary for full 'social' inclusion varies as the means and modes of mobility change and as the potential for 'access' develops with the emergence of new technologies such as charter flights, high speed trains, budget air travel, SUVs, mobile phones, networked computers and so on. These developments transform what is 'necessary' for full.

Moreover, the question of what enables people to move is another aspect largely overlooked in most of the previously mentioned studies. Barriers to movement do not only consist of the disposition of a car or the existence of public transportation in geographic proximity, as will be looked at in the following section.

The Ability to Move and Potential Movement

This section will explore the unequal chances of movement. Thus, mobilities are being linked to other dimensions of inequality beyond the material availability of infrastructures or the black box of individually different preferences and needs After discussing the personal but socially structured preconditions of movement, I will then analyse potential mobility and its link with freedom as an expression of inequality.

As has been pointed out, the increase of technological means has not made spatial position irrelevant. Rather, the ability to move oneself, other people, goods or information has become a powerful force of stratification. In opposition to some views which suggest an understanding of mobility inequality measured as the distances travelled, thereby identifying 'travel poor' (DfT 2000) or 'low mobility groups' (Holden 2007, 6), a high degree of corporeal mobility does not necessarily signify a powerful position within a social space. Rather, as Weiss put it, powerful socio-spatial positions are characterised by a high degree of 'spatial autonomy' consisting of, amongst other things, the availability of advanced technologies of transport and communications and the optimal environment for life and resources (Weiss

2005, 714). This underlines the claim of the mobilities paradigm to analyse the different forms of mobilities together rather than separately, not only to focus on corporeal movement but also on the power to move other people and goods and to include virtual and imaginative movement. It is therefore necessary to differentiate between the compulsion to, and the freedom of, movement. The ability to remain in a suitable geographic place – which is immobility – can also be seen as a privilege and an expression of a powerful social position (Weiss 2005, 714).

In order to untangle the ends of social stratification and movement, Kaufmann (2002) and Kesselring and Vogl (2004) suggest applying the concept of *motility*,[3] which puts the focus of analysis on the question 'what enables people to be mobile' (Kesselring and Vogl 2004, 9). For this purpose, Kaufmann separates out observable movement from motility as the potential for movement, which he defines as 'the way in which an individual appropriates what is possible in the domain of mobility and puts this potential to use for his or her activities' (Kaufmann 2002, 37). Motility as the socially unequal and not randomly distributed ability to be mobile or immobile, as well as to move other people and goods, links dimensions of social inequality, spatial position and material infrastructures with actual movement. These dimensions and links can be found within the three elements which Kaufmann et al. (2004) distinguish as constitutive for motility. These are '*access* to different forms and degrees of mobility, *competence* to recognize and make use of access, and *appropriation* of a particular choice' (Kaufmann et al. 2004, 750, emphasis added). Access refers to the structural aspects, such as the availability of technological means and the conditions of their usage. And, as Urry points out, the most important constraint upon social equality – not only in connection with mobility – are economic resources (Urry 2007, 191). Furthermore, this element refers to spatial position and the relation to spatially unequally distributed infrastructures of connectivity. Competences on the other side include physical abilities, acquired skills and organisational skills which relate to access and appropriation. Thus, this element contains aspects of what Bourdieu (1986) termed cultural capital as well as physical aspects of the body. Finally, appropriation means the perception, interpretation of and acting on options and skills (Kaufmann et al. 2004, 750), which is largely shaped by the social and spatial position and sedimented in the habitus (Bourdieu 2000). Thus, motility rests on a broad range of structural and personal preconditions and resources which represent dimensions of inequality.

---

3   The term *motility* is commonly used in medicine and biology to refer to the ability of either single-celled or multicellular organisms to move spontaneously and independently. Furthermore, it is used in psychoanalysis by Freud and in social theory by Virilio (1998). The earliest usage concerning inequality that I have found, is by Joan Abbott (1966), referring to social mobility.

In a broader approach, Urry develops the notion of *network capital*, which he understands as 'the capacity to engender and sustain social relations' (Urry 2007, 197). Thus, rather than concentrating on personal requirements for movement, he focuses on the relationality of individuals with others which results from the proliferation of new mobilities (Urry 2007, 198). Network capital comprises of eight elements which 'in their combination produce a distinct stratification order that now sits alongside social class, social status and party (Weber 1948, chapter 7)' (Urry 2007, 197). These eight elements are:

1. an array of appropriate documents, visas, money, qualifications that enable safe movement;
2. social contacts at a distance offering hospitality and invitations;
3. physical and informational movement capacities;
4. location-free information and contact points such as real or electronic diaries, answer phones, mobile phones, email etc.;
5. communication devices;
6. appropriate and safe meeting places;
7. access to means of transportation; and
8. time and other resources to manage and coordinate the other seven elements (Urry 2007, 197f).

Each of these elements represents a starting point for further more detailed analysis especially concerning their unequal distribution. For example, Shove (2002) elaborates on the last element, highlighting the stratifying force of time-coordination and organisation skills, which become more and more demanding as the collective socio-temporal order becoming more and more individualised and diverse (Shove 2002, 8).

Whereas Kaufmann et al. (2004) claim that motility can be regarded as a capital in Bourdieu's (1986) sense, since it can be exchanged for other forms of capital, the term of network capital redirects the focus onto the social consequences of mobilities, thus highlighting the co-constitution of social relations and the new scales, forms and interconnectedness of mobilities (Urry 2007, 195). Thus, the ability and potential of movement represents the necessary precondition of any appropriation or mobilisation of resources or capitals, since the latter themselves form specific spatialities (Manderscheid 2008b, 11ff; Allen 2003, 38ff). Acquiring education, transferring cultural capital into an adequate job and thereby into economic capital, sustaining social networks – all these investments into one's potential resources rest on the ability to be mobile which, as has been demonstrated, is closely linked with the disposition of resources. On this line, Urry (2007, 192) and Cass et al. (2005, 543ff) argue that, although mobilities are crucial forces of stratification, social inequality cannot be reduced merely by improving access to the means of mobilities, thus increasing motility. Rather, the broader focus on 'network capital points to the real and potential social relations that mobilities afford'

(Urry 2007, 196). This means that actual and potential movements are not a capital of social value itself but rather a crucial mechanism within the reproduction of inequality.

From a slightly different angle, Sager (2006) points out the importance of the potentiality, rather than just the execution, of movement. Potential mobility is linked with more opportunities and thereby freedom (Sager 2006, 466). Similarly, Bauman highlights the exit option and the possibility to escape into 'sheer inaccessibility' (Bauman 2000, 11) as a powerful element of stratification. This highlights the link between potential mobility as the rejection of territorial confinement (Bauman 2000, 11) and the libertarian idea of freedom as self-determination, understood as 'the right or opportunity of individuals to make choices so as to be in charge of their own fates' (Sager 2006, 466). Crucial in this context is the individual autonomy to decide whether or not to act on the possibilities or opportunities, which also involves the choice between travelling and not travelling (Sager 2006, 469). But, as has been argued, the potential to be mobile seems to vary with other dimensions of inequality such as economic and cultural capital, spatial position or incorporated schemes of perception. Thus, a lack of motility and network capital does not only constrain one's chance to appropriate socially valued goods and resources but also represents an experienced constraint of freedom as an expression of lived inequality relations in a social world of increasing spatially dispersed relations. On the other hand and as a mirror-image of the *ability* to move, the *compulsion* to move, e.g. forced labour migration, also constitutes an experienced constraint of freedom since it may weaken or cut important social relations through unwilled distanciation from one's personal meaningful social networks.[4]

Sager's reflection on the link between potential movement and freedom points at another aspect in connection with the discussion of unequal mobilities. The approaches so far discussed see the reasons for movement in access to activities and goods and thus as part of social participation. However, the meaning of mobilities goes further than their endpoints and the corresponding social and cultural embedding of movement is highly contested and stratified.

Social Meaning of Movement

> What matters is not just 'who can travel where, when and how?' but also 'who gets to tell the story?' (Frello 2008, 32)

In contrast to the understanding of movement as a derived demand which is found in most transportation studies, the mobilities turn emphasises its socio-cultural embedding. That means that the meaning of movement is socially constituted and thus structured by power relations.

---

4 I want to thank Noel Cass for emphasising this point.

The notion of movement and its promise of progress is inextricably linked with capitalist modernity (cf. Rammler 2008). Overcoming spatial distances and moving beyond borders was described by Marx as a fundamental characteristic of capital (Marx 1973 [1861], 524). Together with capitalist modernity's ongoing processes of differentiation and separation, the spatial extension of elongated chains of action increased (Elias 1999) which contributes to an increase of geographical mobilities.[5]

Yet, this continuing growth in transport is not just a derived function of the capitalist extension of markets or growing geographic spreading of social relations. Rather, certain ways and forms of movement provide symbolic profit beyond the mere access of valued locations, activities or goods. Elaborating on the social constitution of meaning of mobilities and 'the geographical imaginations that lie behind mobilisation in a diverse array of contexts', Cresswell (2006, 2) suggests we make an analytical distinction between movement and mobility: movement describes the idea of an act of displacement that allows people to move between locations, thus a general fact before the type, strategies and its social implications are considered. Mobility, on the other hand, refers to the socially produced motion which is understood through three relational moments,[6] that is, first, mobility as a brute fact, potentially observable and closest to pure motion. Second, mobility can be conveyed through representational strategies such as film or law, medicine or literature. These representations make sense of the movement through the production of meanings that tend to be ideological. Third, human mobility refers to an irreducible embodied experience; it is a way of being in the world (Cresswell 2006, 2–4).

Cresswell concentrates on the representational strategies and argues that the interweaving of modernity and mobility can be read through representations and mobile metaphors, for example, the train journey as a metaphor for a specific kind of modernity, Walter Benjamin's *flâneur* strolling the city unbound and free representing the modern urban tension of physical proximity and social distance (Cresswell 2006, 19; Urry 2007, 69ff; 90ff), or tourists, vagrants and pilgrims (Bauman 1996; Urry 2000, 26ff). As a product of power relations, these continuously contested meanings and representations of mobilities contain ambiguities, social hierarchies and exclusions. Thus, analyses of the contingent discursive constitution of the notions of progressive and destructive mobilities as well as the differentiation of mobility and immobility are another

---

5 Although the world's population grew during the previous century by a factor of four, motorised passenger kilometres and tonne-kilometres grew on average by a factor of about 100 by all modes each (Holden 2007, 3; OECD 2000).

6 The three relational moments of Cresswell's mobilities run largely parallel to Lefebvre's (1991, 38) trialectic of space as lived space, conceived space (representations of space) and perceived space (spatial practice).

way to tackle the question of the interconnectedness of social structures and movements which appear deeply infused with aspects of power.

The perspective on mobility and what associations are evoked, depends largely on who moves. A negative association of mobilities appears for example in the reference to 'mob' as the mobilised crowd, which is associated with chaos rather than freedom (Urry 2007, 205; Kaplan 2006, 396). Similarly, immigrants from outside Europe are commonly referred to as 'flooding in' whereas western highly qualified migrants are valued as 'expatriates'. As Urry phrases it, 'states routinely hold that there are good movers and bad movers and that the latter should be limited, penalised, extradited or thrown into prison' (Urry 2007, 205). The link of modernity and mobility thus has not always been perceived as positive and progressive; rather, there has been also an accompanying discourse concerning destructive forces of movement. This discursive line can be traced through history – for example, within the narrative of anti-urbanism – and seems deeply rooted in some socially embodied 'metaphysics of sedentarism', which links people to places and territories (Cattan 2008, 85). Contemporary expressions can be found in Putnam's (1995) assumption of a decrease of social capital operationalised as local community engagement as well as in Sennett's (1998) critique of flexible capitalism seen as incompatible with the development of personal stable characters. In these views, movement is perceived as a fundamental threat to the spatial and social cohesion of territories (Cattan 2008, 85) as well as of coherent personalities.

The focus on mobility as foundation of modern and globalised worlds involves an ontological critique of these sedentarist metaphysics and seemingly fixed categories, such as 'society', 'nation', 'neighbourhood'. These were then replaced with more mobile terms such as 'flows' or 'scapes' (Cresswell 2006, 25ff; Frello 2008, 26; Urry 2007, 31ff). Yet, the positive view of mobilities remains within a hierarchical binary scheme of the mobile and the immobile and runs the risk of turning into a 'nomadic metaphysics' (Cresswell 2006, 42–54) which is by no means more 'objective' but also strongly shaped by social power relations. As Kaplan (2006, 395) has pointed out, the tendency in social theory to romanticise mobility as a free-floating alternative to the rooted traditions of place. But, after everything has been mobilised, the notion of mobility will lose all analytical power. Therefore, mobility and immobility should be understood as profoundly relational and experiential, movement and standstill may be carried out and experienced in many different ways depending on the context (Adey 2006, 83). And these relations appear to be deeply interwoven with relations of class, ethnicity, age and gender.

Some trailblazing work has been carried out, especially concerning the gendered notion of mobilities and its static counterpart with their connected metaphors and imaginations. Amongst others, Wolff (1993, 230) argues that there is an 'intrinsic relationship between masculinity and travel' which, although it never stopped (some) women travelling, runs very deep through the socio-spatial order of modernity. Thus, and by applying a relational understanding of the mobile/immobile dualism (Kaplan 2006, 406), mobility seems to be connected with masculinity, whereas sessility or sedentarism appears as feminine[7] (Wolff 1993, 229). These gendered constructions are verbalised for example when 'men go to war' for the sake of 'women and children at home', as Enloe (2004) points out, and also in the more general and close link of home, tradition and reproduction as the feminine sphere (Yuval-Davis 2003, 313ff).[8]

Also challenging the power infused dualism of the mobile and the static, Frello (2008) highlights how certain activities are not recognised as mobilities due to lack of power of defining.

> To describe an activity in terms of motion is to declare that it implies the transgression of a difference. Hence it also involves the declaration of the relevance of certain types of distinction, such as the distinction between 'here' and 'there' or between different academic disciplines or different imaginary worlds. It is the distinction that constitutes the difference that is overcome by 'movement'. (Frello 2008, 32)

This discursive approach on movement deconstructs what Cresswell (2006, 3) calls the 'general fact of displacement' as something preceding its social discursive embedding. This enables Frello to elucidate how certain forms of mobilities are being refused the status of overcoming a difference in order to perpetuate an exclusive social order. The social discursive construction and recognition of this difference as significant represents the precondition for movement to be recognised as movement, that is, as overcoming some type of distance (Frello 2008, 32). Frello illustrates this with the case of second generation immigrants who had been 'sent home' for committing a crime in their 'host-country'. Thereby, the immigration of their parents is socially not recognised as overcoming a distance and becoming part of a new society and even their offspring are denied the status of citizens. In contrast, international moves of the global elites are commonly recognised as broadening experience and cosmopolitanism, thus overcoming a significant difference.[9] Here, mobility is a metaphor for various activities and attitudes, such as imagination and openness towards other ways of thinking and living (Frello 2008, 28; Cresswell 2006, 2f).

---

7   Wolff (1993, 230) explicitly understands the gendered notion of mobility as intrinsic rather than essentialist.

8   This gendered notion of urban space has been studied by Frank (2003) or Wilson (1991) amongst others and carries further a notion of mobility (the *flâneur*) and immobility (the home). A recent excellent collection of more work on gender and mobilities is found in Uteng and Cresswell (2008).

9   This can be found, for example, for academics, when leaving their country of origin and working in another country. It is linked with the symbolic profit of being 'international' and considered a crucial part of an academic profile.

These examples highlight the social context of power relations which infuses the cultural production of meaning of mobilities. The discursive embedding of movement thus privileges some and disregards other movers and mobilities, thereby impacting their social position. Thus, in order to understand the elements of power and inequality contained in these collections of discourses, they should be analysed by asking who shapes the debate, how they shape it and who gains support from which side. Moreover, when analysing transportation policies and the corresponding discursive production of meanings which largely reduce mobilities to car usage and the time-space organisation of employed men, the inherent excluding force, which largely ignores the spatio-temporalities of children, homemakers and old people outside this dominant culture (Hamilton 2003; Peters 2006, 128ff), is unveiled. The link between these sense-making narratives, material landscapes and practices of movement will be discussed in the following section.

Practices of Mobility and Passages

The fourth focus on the interweaving of mobilities and inequalities is also the most complex one, as it extends the view onto the emergence of social practices and in doing so refers back to the other three. The way in which people and social groups react to and interact with material networked landscapes of mobilities and technologies of transportation and communication is often conceptualised in either a social or technology deterministic way. Furthermore, the analyses of this interplay often ignore its stratified and stratifying aspects. The following will outline some promising points of departure for further research which is able to understand practices of movement and their relation to the material infrastructure in a more social theory informed manner than the studies of actual movement mentioned at the beginning of the paper.

Concerning the relation of communication and transportation technologies on one side and social relations and practices on the other, there are two main shortcomings found in both the public and academic debate. The first socio-deterministic one gives primacy to social relations, as Wajcman (2008, 66) phrases it, treating them as existing prior to and outside the technology, which then allows people to do the same things as before, but faster. This argument runs through the history of transport planning which tries to solve traffic problems by the provision of broader roads (cf. Peters 2006, 9ff) as well as through most transportation studies with their rooted concepts of mobility exclusion which understand movement as a derived demand from a predefined set of needs (see above). This becomes problematic, since analyses based on this argument fail to realise that social practices, social relations and the connected question of what is necessary for social participation vary and change as the means and modes of mobilities change (Cass et al. 2005, 542). 'New technologies reconfigure relationships between people and the spaces they occupy, altering the basis of social interaction' (Wajcman 2008, 66). This becomes apparent, for example, when faced with the paradox that although technologies have become ever faster, people feel more and more rushed rather than having more spare time (Rosa 2005, 135ff; Shove 2002; Southerton 2003; Wajcman 2008).

The second shortcoming consists somehow of the opposite, that is, assuming a direct causal link between the emergence of communication and transportation technologies and a specific form of usage: This technological-deterministic argument underlies large parts of social theory which explains processes of globalisation, the increase of social relations over large geographic distances and an increase of travelled distances as a function of the emergence of new and faster technologies of transportation and communication.[10] Yet, as argued by an increasing number of scholars, the usage and appropriation of these means cannot be deduced from the technological potential; rather, technological artefacts should be conceived of as culturally and socially situated (Wajcman 2008, 67). This means that the way in which technologies are integrated into the social world and become part of social practices is contingent and an empirical question. Further, the 'same technologies can mean very different things to different groups of people, collectively producing new patterns of social interaction, new relationships, new identities' (Wajcman 2008, 70). For example mobile phones are transforming practices of meeting from fixed to fluid and constantly re-negotiable time-spaces, as Larsen et al. (2006) have shown. This has not been a predefined practice inherent in the technology itself. In order to participate in these fluid time-spaces of social relations, it is not necessary only to have a mobile phone, as these practices require specific skills, especially the ability to juggle individualised and more complex time schedules (Shove 2002). As has been argued already in connection with the concepts of motility and network capital, these skills and abilities are not equally distributed and thus form a increasingly powerful stratifying force. That means that dominant practices emerge within social power relations and may exclude, block or override more traditional, alternative or marginal practices.

Referring to this complex interweaving of social and technological relations, Peters succeeds in overcoming the widespread technological deterministic arguments by suggesting the broader concept of passages which links the emergence of technologies and the production of meaning and material landscapes with social practices, against the background of social power relations. Using the example of Thomas Cook's innovations in Britain in the middle of the nineteenth century as an actor who organised travel as a complex spatio-temporal order of heterogeneous elements, accessible to a rather broad

---

10   Besides its techno-determinism, this argument abstracts from the networked structure of these technologies, which are not at all ubiquitous (cf. Graham and Marvin 2001).

part of society, Peters exemplifies how, from an actor's perspective, technologies are actively embedded into socio-material settings (Peters 2006, 50–72). What seems *ex post* as a self-evident innovation – organised package holidays – was actively constructed by combining several heterogeneous elements:

> New means of transportation were not a sufficient precondition for Cook to offer his customers fast journeys. Enough travellers had to be mobilized to keep prices low; negotiations about the tariffs offered by different rail companies had to be made; arrangements had to be made with hotel and restaurant owners; place myths had to be invented; there had to be travel guides in which Cook's destinations were described; and Cook had to place employees all along the route to solve problems, making it possible for his passengers to continue travelling. As situated relations between time and space, the passages of Cook imply an order which contained material as well as immaterial entities. ... The passages of Cook made it possible to travel faster, but not for everyone, and not to every imaginable place or at every moment. (Peters 2006, 72)

Thus, these passages contain a social and spatial excluding moment which contradicts the assumption that everyone could go everywhere at every time. By linking movement with discursively-produced meaning and material landscapes with the concept of passages, Peters uncovers the politically-produced path of transportation as one amongst other possible paths of ordering socio-spatial complexity:

> In order to travel, I claim, we need to construct passages that produce a situated relation between time and space. ... As heterogeneous orders, passages assume both material and discursive elements. As planned yet contingent orders, they must be 'repaired' continuously in real time. And as orders that both include and exclude people, places and moments in time, they are inherently political and have to be justified and legitimated. (Peters 2006, 2)

On the discursive level, a specific ensemble of ideas, concepts and categorisations gives meaning to physical and social realities (Peters 2006, 18). Moreover, these power-infused discursive meanings themselves shape material (and social) space since they set the frame of spatial policies which in turn has consequences for the interrelations with other dimensions of inequality (Richardson and Jensen 2008, 220). For example, with the emergence of the car as the dominant means of transportation, the sense-making narrative produced specific corresponding landscapes of roads and car-centred cities. Together with a whole set of infrastructures like petrol stations and hotel accommodation, the car culture shaped a very specific practice of travelling as another passage, which Peters (2006, 73–99) illustrates. More generally, the 'system of automobility' (Urry 2004) and its discursive link to freedom and independence has fundamentally impacted on urban space and contributed to suburban life and its car centrism as a still powerful ideal of the good family life (Freudendal-Pedersen 2007). Making the link back to practices, it can be argued that these socially-produced discourses and material landscapes are sedimented and ingrained in people's pre-reflexive perception of the world or their habitus,[11] thus forming some form of incorporated passages which pre-structure socio-spatial practices. Thus, spatial practices appear as shaped by socio-political and thereby stratified forces rather than originating from predefined demands or a black box of individual needs A similar argument is applied by Freudendal-Pedersen, who found in her research *structural stories* as the communicated expression of the internalised socio-material formations:

> A structural story contains the arguments people commonly use to explain their actions and decisions. A structural story is used to explain the rationalities behind the way we act and the choices we make when exercising our daily routines and is a guide to certain actions. The structural stories form the basis that determines how the individual views certain problems and their solutions. The social practice of the individual produces and reproduces these structural stories. (Freudendal-Pedersen 2007, 29)

Structural stories, for example, concerning the car, guide and rationalise choices in favour of a car centred, often suburban lifestyle and they cut alternatives out of perception. Thus, structural stories represent dominant sense making discourses which stabilise the dominant mobility patterns and practices and hinder others. Yet, as has been argued above, these passages and their inherent mobilities rest on unequally available prerequisites, thereby limiting the chance to participate in these 'normal' practices. Life arrangements and mobility patterns outside these dominant passages cannot rely on the same degree of social acceptance and normality (Shove 2002, 2).

Conclusion

The point of departure for this chapter was the assumption that mobilities and social inequalities are complexly interwoven, an assumption which is not yet sufficiently explored and underpinned with theory-informed

---

11   Bourdieu (2000) conceptualises the *habitus* as incorporated social structures, including schemes of perception and a sense for one's place, which generate patterns of practices adequate to one's social position. Adapted for a space- and mobility-sensitive analysis of inequalities, sedimenting the social distinctions that separate individuals in time and space the habitus generates practices adequate to the socio-spatial position occupied (cf. Fiedland and Boden 1994, 22; Manderscheid 2008a).

research. The capacity to move seems to have gained more significance as a crucial mechanism in the reproduction of the socio-spatial order. Thus, the importance of this research endeavour is not a mere academic one but stems also from the ecological limit to increased motorised traffic and the search for alternative futures and political solutions. As with an increased taxation, very often applied means to reduce motorised transport are in conflict with issues of social equity and justice. Imagining possible futures should therefore be grounded in an understanding of these complex socio-material configurations of mobilities. By the same token, Urry argues that:

> ... futures are heavily circumscribed and are clearly not at all open. Some of the key determinants of such futures include cognitive and non-cognitive human capacities, the embedded traditions within each society, the power and conserving effect of national and international states, huge global processes operating on multiple levels, the relative fixity of the built environment, various economic, technological and social path dependencies, and large-scale enduring economic-technological, social, environmental and political inequalities around the globe. Thus there are very powerful socio-physical systems, moving in and through different time-spaces, and these produce many constraints upon possible 'futures'. There are countless ways in which different levels and orders are locked in and limit the possibilities of future change. (Urry 2007, 276)

As a conceptual framework for analysing the interweaving of mobilities and inequalities, I discussed four perspectives, those of actual movement based on networked infrastructures, the ability and potential to move, the social meaning and practices of movement. These approaches are combined within the concept of passages as socio-spatial arrangements.

Looking at the material infrastructures of movement and communication, I briefly outlined a paradigm change from a national-territorial ideal to the dominance of networked spaces. By this I refer to processes of spatial polarisation: on one hand, already highly central and connected places and social elites are becoming continuously more connected through infrastructures and benefit largely from corresponding social, political and economic attention. On the other hand, and sometimes in close geographic proximity, new peripheries of bypassed regions and social groups emerge.

Large parts of transportation research in this vein of unequal mobility seem preoccupied with a rather descriptive and quantitative inventory of the correlation of movement, transportation and other socio-demographic variables of inequality which provides the necessary ground for further studies. Yet, from a more sociologically informed view, the underlying rational choice approach in particular fails to capture how the ability to move as well as the social meaning of actual movement are shaped. Further, reducing mobilities to time-wasting movement in geographic space underestimates the cultural embeddings of technologies of transportation and communication. Therefore, these conceptualisations of unequal mobilities tend to overrate the chances to change mobility patterns by means of policies and planning.

Broadening the view to the unequal abilities of movement, the discussion of the concepts of motility and network capital demonstrated the close link to unequally available resources and skills. Thus, the material infrastructural base of mobilities represents only one amongst many elements which enable people to move. Yet, as the section on the constitution of meaning revealed, movement is embedded within a socio-spatial context of sense-making discourses, which are structured by social power relations. That means that not all activities are perceived as meaningful movement; rather, certain activities are linked with symbolic profit while others are not recognised as overcoming fractions or distances at all. Against the background of continuously increasing elongations of chains of action in a modernising and globalising world, mobilities and their socially contested significance have become a increasingly relevant field within the social space of power relations and positions (Bourdieu 2000).

Although the discursive embedding of movement was unveiled as contingent, thus suggesting possibilities for alternative mobility narratives, the last section of the chapter suggested that in order to change existing mobility practices and create new ones, more complex and radical re-orderings of technologies, material landscapes and discourses into new passages have to take place. The fundamental interrelation of (capitalist) modernity and globalisation and the accompanying spatial stretching of social relations challenges the aim of sustainability politics to reduce transportation of goods and people, since the significance of mobilities as a precondition of social participation appears impossible to turn back (Rammler 2008, 72). Thus, the ecological necessity to reduce motorised movement is not only a question of people's values or the built environment but challenges the heart of capitalist socio-economic formations. The significance of mobility as the basis of social participation has increased with the spatial stretching of social relations. Maybe, against the background of the climate change and spatial polarisation, there is a new social question emerging which is not merely about the distribution of material wealth and the position within the relations of production, but is also about the distribution of increasingly scarce network capital and the position within the spatial relations of power.

References

Abbott, J. (1966), 'The Concept of Motility', *The Sociological Review* 14:2, 153–61.
Adey, P. (2006), 'If Mobility is Everything then it is Nothing: Towards a Relational Politics of (Im)Mobilities', *Mobilities* 1:1, 75–94.

Bauman, Z. (1996), 'From Pilgrim to Tourist – or a Short History of Identity', in Hall, S. and du Gay, P. (eds), *Questions of Cultural Identity* (London: Sage), 18–36.
Bauman, Z. (2000), *Liquid Modernity* (Cambridge: Polity Press).
Beck, U. (2007), 'Beyond Class and Nation: Reframing Social Inequalities in a Globalizing World', *The British Journal of Sociology* 58:4, 679–705.
Black, W.R. and Nijkamp, P. (eds) (2002), *Social Change and Sustainable Transport* (Bloomington, IN: Indiana University Press).
Bourdieu, P. (1986), 'The (Three) Forms of Capital', in Richardson, J.G. (ed.), *Handbook of Theory and Research in the Sociology of Education* (Westport, CT: Greenwood Press), 241–58.
Bourdieu, P. (2000), *Distinction. A Social Critique of the Judgment of Taste* (Cambridge, MA: Harvard University Press).
Bradley, H. (1996), *Fractured Identities. Changing Patterns of Inequality* (Cambridge: Polity Press).
Bundesamt für Raumentwicklung (ARE) (2006), *Raumstruktur und Mobilität von Personen. Ergebnisse einer Sonderauswertung des Mikrozensus 2000 zum Verkehrsverhalten* (Bern).
Canzler, W., Kaufmann, V. and Kesselring, S. (eds) (2008), *Tracing Mobilities: Towards a Cosmopolitan Perspective* (Aldershot: Ashgate).
Cass, N., Shove, E. and Urry, J. (2005), 'Social Exclusion, Mobility and Access', *Sociological Review* 53:3, 539–55.
Castells, M. (2001), *Der Aufstieg der Netzwerkgesellschaft* (Opladen: Leske + Budrich).
Castells, M. (2002 [1996]), 'The Space of Flows', in Susser, I. (ed.), *The Castells Reader on Cities and Social Theory* (Oxford: Blackwell), 390–406.
Cattan, N. (2008), 'Gendering Mobility: Insights into the Construction of Spatial Concepts', in Uteng, T.P. and Cresswell, T. (eds).
Cresswell, T. (2006), *On the Move: Mobility in the Modern Western World* (London: Routledge).
Cresswell, T. and Uteng, T.P. (2008), 'Gendered Mobilities: Towards a Holistic Understanding', in Uteng, T.P. and Cresswell, T. (eds), *Gendered Mobilities* (Aldershot: Ashgate).
Department for Transport (DfT) (2000), *Social Exclusion and the Provision and Availability of Public Transport Report* (London).
Drewes Nielsen, L. (2005), 'Reflexive Mobility – A Critical and Action Oriented Perspective on Transport Research', in Thomsen, T.U., Drewes Nielsen, L. and Gudmundsson, H. (eds), *Social Perspectives on Mobility* (Aldershot: Ashgate), 47–64.
Elias, N. (1999 [1939]), *Über den Prozess der Zivilisation. Soziogenetische und psychogenetische Untersuchungen. Zweiter Band Wandlungen der Gesellschaft. Entwurf zu einer Theorie der Zivilisation* (Frankfurt/M: Suhrkamp).

Enloe, C. (2004), *The Curious Feminist: Searching for Women in a New Age of Empire* (Berkeley, CA: University of California Press).

Erikson, R. and Goldthorpe, J.H. (1992), *The Constant Flux. A Study of Class Mobility in Industrial Societies* (Oxford: Clarendon Press).

Flade, A. and Limbourg, M. (eds) (1999), *Frauen und Männer in der mobilen Gesellschaft* (Opladen: Leske + Budrich).

Frank, S. (2003), *Stadtplanung im Geschlechterkampf. Stadt und Geschlecht in der Großstadtentwicklung des 19. und 20. Jahrhunderts* (Opladen: Leske + Budrich).

Fraser, N. (2003), 'Soziale Gerechtigkeit im Zeitalter der Identitätspolitik. Umverteilung, Anerkennung und Beteiligung', in Fraser, N. and Honneth, A. (eds), *Umverteilung oder Anerkennung? Eine politisch-philosophische Kontroverse* (Frankfurt/Main: Suhrkamp), 13–128.

Frello, B. (2008), 'Towards a Discursive Analytics of Movment: On the Making and Unmaking of Movement as an Object of Knowledge', *Mobilities* 3:1, 25–50.

Freudendal-Pedersen, M. (2007), 'Mobility, Motility and Freedom: The Structual Story as an Analytical Tool for Understanding the Interconnection', *Schweizerische Zeitschrift für Soziologie* 33:1, 27–43.

Friedland, R. and Boden, D. (1994), 'Nowhere: An Introduction to Space, Time and Modernity', in Friedland, R. and Boden, D. (eds), *Nowhere: An Introduction to Space, Time and Modernity* (Berkeley, CA: University of California Press), 1–60.

Götz, K. Loose, W., Schmied, M. and Schubert S. (2003), *Mobilitätsstile in der Freizeit. Minderung der Umweltbelastungen des Freizeit- und Tourismusverkehrs* (Berlin: Erich Schmidt Verlag).

Günzel, S. (2007), 'Raum – Topographie – Topologie', in Günzel, S. (ed.), *Topologie* (Bielefeld: transcript).

Graham, S. (2002), 'Flowcity: Networked Mobilities and the Contemporary Metropolis', *Journal of Urban Technology* 9:1, 1–20.

Graham, S. and Marvin, S. (2001), *Splintering Urbanism. Networked Infrastructures, Technological Mobilities and the Urban Condition* (Oxon: Routledge).

Hamilton, K. (2003), 'If Public Transport is the Answer, What is the Question?' in Root, A. (ed.), *Delivering Sustainable Transport – A Social Science Perspective* (Oxford: Pergamon).

Harvey, D. (1990), *The Condition of Postmodernity. An Enquiry into the Origins of Cultural Change* (London: Blackwell).

Healey, P. (2004), 'The Treatment of Space and Place in the New Strategic Spatial Planning in Europe', *International Journal of Urban and Regional Research* 28:1, 45–67.

Hine, J. and Mitchell, F. (2003), *Transport Disadvantage and Social Exclusion: Exclusionary Mechanisms in Transport in Urban Scotland* (Aldershot: Ashgate).

Holden, E. (2007), *Achieving Sustainable Mobility. Everyday and Leisure-Time Travel in the EU* (Alsershot: Ashgate).

Jetzkowitz, J., Schneider, J. and Brunzel, S. (2007), 'Suburbanisation, Mobility and the 'Good Life in the Country': A Lifestyle Approach to the Sociology of Urban Sprawl in Germany', *Sociologia Ruralis* 47:2, 148–71.

Jiron, P.M. (2007), 'Unravelling Invisible Inequalities in the City through Urban Daily Mobility. The Case of Santiago De Chile', *Schweizerische Zeitschrift für Soziologie* 33:1, 45–68.

Kaplan, C. (2006), 'Mobility and War: The Cosmic View of US 'Air Power'', *Environment and Planning A* 38:2, 395–407.

Kaufmann, V. (2002), *Re-Thinking Mobility: Contemporary Sociology* (Aldershot: Ashgate).

Kaufmann, V., Bergman, M.M. and Joye, D. (2004), 'Motility: Mobility as Capital', *International Journal of Urban and Regional Research* 28:4, 745–65.

Kessel, F. and Reutlinger, C. (eds) (2008), *Schlüsselwerke der Sozialraumforschung* (Wiesbaden: VS Verlag für Sozialwissenschaften).

Kesselring, S. and Vogl, G. (2004), 'Mobility Pioneers. Networks, Scapes and Flows between First and Second modernity', paper presented at Mobility and the Cosmopolitan Perspective Workshop, 30 January, München.

Larsen, J., Axhausen, K.W. and Urry, J. (2006), 'Geographies of Social Networks: Meetings, Travel and Communications', *Mobilities* 1:2, 261–83.

Löw, M. (2001), *Raumsoziologie* (Frankfurt/M.: Suhrkamp).

Manderscheid, K. (2004), *Milieu, Urbanität und Raum. Soziale Prägung und Wirkung städtebaulich gebauter Räume* (Wiesbaden: VS Verlag für Sozialwissenschaften).

Manderscheid, K. (2008a), 'Pierre Bourdieu – Ein ungleichheitstheoretischer Zugang zur Sozialraum-Analyse', in Kessel, F. and Reutlinger, C. (eds), *Schlüsselwerke der Sozialraumforschung* (Wiesbaden: VS Verlag für Sozialwissenschaften).

Manderscheid, K. (2008b), 'Integrating Space and Mobilities into the Analysis of Social Inequality', paper presented at CeMoRe Seminar, 22 January 2008, Lancaster.

Manderscheid, K. and Bergman, M.M. (2008), 'Spatial Patterns and Social Inequality in Switzerland – Modern or Postmodern?', in Pflieger, G., Pattaroni, L., Jemelin, C. and Kaufmann, V. (eds), *The Social Fabric of the Networked City* (London: Routledge), 41–65.

Marx, K. (1973 [1861]), *Grundrisse. Outlines of the Critique of Political Economy* (Harmondsworth: Penguin).

Massey, D. (2005), *For Space* (London: Thousand Oaks).

McKenzie, C. (2003), 'Transport, Modernity and Globalisation', in Root, A. (ed.), *Delivering Sustainable Transport – A Social Science Perspective* (Oxford: Pergamon).

Næss, P. (2006), *Urban Structure Matters. Residential Location, Car Dependence and Travel Behaviour* (London: Routledge).
Næss, P. (2008), 'Gender Differences in the Influences of Urban Structure on Daily Travel', in Uteng, T.P. and Cresswell, T. (eds), *Gendered Mobilities* (Aldershot: Ashgate).
Noller, P. (1999), *Globalisierung, Stadträume und Lebensstile. Kulturelle und lokale Repräsentationen des globalen Raumes* (Opladen: Leske + Budrich).
Nowicka, M. (2006), 'Mobility, Space and Social Structuration in the Second Modernity and Beyond', *Mobilities* 1:3, 411–35.
OECD (2000), 'Environmentally Sustainable Transport. Futures, Strategies and Best Practices. Synthesis Report of the OECD Project on Environmentally Sustainable Transport EST', presented on the Occasion of the International Est! Conference, 4–6 October, Vienna.
OECD (2006), *Decoupling the Environmental Impacts of Transport from Economic Growth* (Vienna: OECD Publishing).
Peters, P. (2006), *Time, Innovation and Mobilities. Travel in Technological Cultures* (London: Routledge).
Pflieger, G., Pattaroni, L., Jemelin, C. and Kaufmann, V. (eds) (2008), *The Social Fabric of the Networked City* (London: Routledge).
Putnam, R.D. (1995), 'Bowling Alone: America's Declining Social Capital', *Journal of Democracy* 6:1, 65–78.
Rammler, S. (2008), 'The Wahlverwandtschaft of Modernity and Mobility', in Canzler, W., Kaufmann, V. and Kesselring, S. (eds), *Tracing Mobilities: Towards a Cosmopolitan Perspective* (Aldershot: Ashgate), 57–76.
Richardson, J.G. (ed.) (1986). *Handbook of Theory and Research in the Sociology of Education* (New York: Greenwood Press).
Richardson, J.G. (2006), 'The Thin Simplification of European Space: Dangerous Calculations?' *Comparative European Politics* 4:1, 203–17.
Richardson, T. and Jensen, O.B. (2003), 'Linking Discourse and Space: Towards a Cultural Sociology of Space in Analysing Spatial Policy Discourses', *Urban Studies* 40:1, 7–22.
Ronneberger, K., Lanz, S. and Jahn, W. (1999), *Die Stadt als Beute* (Bonn: Dietz).
Root, A., Button, K.J. and Schintler, L. (2002), 'Women and Travel. The Sustainability Implications of Changing Roles', in Black, W.R. and Nijkamp, P. (eds) (2002), *Social Change and Sustainable Transport* (Bloomington, IN: Indiana University Press), 149–56.
Rosa, H. (2005), *Beschleunigung. Die Veränderung der Zeitstrukturen in der Moderne* (Frankfurt/Main: Suhrkamp).
Sager, T. (2006), 'Freedom as Mobility: Implications of the Distinction between Actual and Potential Travelling', *Mobilities* 1:3, 465–88.
Sassen, S. (2001), *The Global City* (Princeton, NJ: Princeton University Press).

Scheiner, J. and Kasper, B. (2003), 'Lifestyles, Choice of Housing Location and Daily Mobility: The Lifestyle Approach in the Context of Spatial Mobility and Planning', *International Social Science Journal* 55:2, 319–32.

Schulze, G. (1992), *Die Erlebnisgesellschaft. Kultursoziologie der Gegenwart* (Frankfurt am Main: Campus Verlag).

Sennett, R. (1998), *The Corrosion of Character: The Personal Consequences of Work in the New Capitalism* (New York: Norton).

Sheller, M. and Urry, J. (2006), 'The New Mobilities Paradigm', *Environment and Planning A* 38:2, 207–26.

Shields, R. (1992), 'A Truant Proximity: Presence and Absence in the Space of Modernity', *Environment and Planning D* 10:2, 181–98.

Shove, E. (2003), 'Rushing Around: Coordination, Mobility and Inequality', Department of Sociology, Lancaster University, (home page) http://www.comp.lancs.ac.uk/sociology/papers/Shove-Rushing-Around.pdf, accessed 27 May 2008.

Southerton, D. (2003), "Squeezing Time': Allocating Practices, Coordinating Networks and Scheduling Society', *Time Society* 12:1, 5–25.

Tienda, M. and Grusky, D.B. (eds) (1994), *Social Stratification. Class, Race, and Gender in Sociological Perspective* (Boulder, CO: Westview Press).

Thomsen, T.U., Drewes Nielsen, L. and Gudmundsson, H. (eds) (2005), *Social Perspectives on Mobility* (Aldershot: Ashgate).

Urry, J. (2000), *Sociology beyond Societies: Mobilities for the Twenty-First Century* (New York: Routledge).

Urry, J. (2004), 'The 'System' of Automobility', *Theory, Culture and Society* 21:4/5, 25–39.

Urry, J. (2007), *Mobilities* (Cambridge: Polity).

Uteng, T. P. and Cresswell, T. (eds) (2008), *Gendered Mobilities* (Aldershot: Ashgate).

Virilio, P. (1998), *Rasender Stillstand. Essay* (Frankfurt am Main: Fischer Taschenbuch Verlag).

Wajcman, J. (2008), 'Life in the Fast Lane? Towards a Sociology of Technology and Time', *The British Journal of Sociology* 59:1, 59–77.

Walby, S. (2001), 'From Community to Coalition: The Politics of Recognition as the Handmaiden of the Politics of Redistribution', *Theory, Culture and Society* 18:2, 113–35.

Weiss, A. (2005), 'The Transnationalization of Social Inequality: Conceptualizing Social Positions on a World Scale', *Current Sociology* 53:4, 707–28.

Wilson, E. (1991), *The Sphinx in the City. Urban Life, the Control of Disorder, and Women* (Berkeley, CA: University of California Press).

Wolff, J. (1993), 'On the Road Again: Metaphors of Travel in Cultural Criticism', *Cultural Studies* 7:2, 224–39.

Chapter 3
Life Course Inequalities in the Globalisation Process

Hans-Peter Blossfeld, Sandra Buchholz and Dirk Hofäcker

Introduction

Over the last two decades, globalisation has had a lasting impact on modern societies and globalisation research now plays an important role in the social and economic science discourse. However, it is notable that, up to now, this research mostly has been limited to describing and analysing the effects of the globalisation process on national aggregates with the help of macro data (see, e.g., Panic 2003). Although there is no denying that this has made a major contribution to the understanding of globalisation, it can be assumed that there has been an equally fundamental influence on the micro level, particularly on the development of individual life courses in modern societies as the spatial dynamics caused by globalisation – for example, the rapid worldwide networking through new information and communication technologies – have strongly affected the mobility of labour and capital and, as a result, the (re-)structuring of social inequalities.

The GLOBALIFE[1] research project goal was to perform a cross-nationally comparative empirical analysis of the effects of the globalisation process on the life-courses of women and men in advanced industrial societies. In a sequence of four project phases, 71 social scientists compared the effects of the globalisation process on selected transitions in the individual life course in a total of 17 OECD member states. One major purpose of these analyses was to use current longitudinal datasets and advanced longitudinal analysis to work out how far national institutional structures 'filter' the globalisation process in specific ways, thus producing outcomes on the level of individual life courses that differ from country to country and that act to produce certain types of social inequalities within them.

This chapter summarises the key findings of the GLOBALIFE project. It starts from a theoretical perspective by sketching the characteristics of the globalisation process and its effects on individual life courses. These theoretical

---

1　The GLOBALIFE research project, supported by a €1.7 million grant from the Volkswagen Foundation in Germany, was carried out at the Universities of Bielefeld and Bamberg from 1999 to 2006.

considerations are then confronted with the most important empirical findings of the GLOBALIFE project which studied the *general changes* in central transitions of the life course resulting from globalisation, but also the *country-specific form* that these changes take in various modern societies.

Features and Effects of the Globalisation Process

Basically, globalisation is understood as a combination of processes that have led to growing worldwide interconnectedness (Alasuutari 2000; Robertson 1992). While this is certainly no fundamentally new phenomenon (Alasuutari 2000; Robertson 1990; Sutcliffe and Glyn 1999), nobody would deny that most industrial nations have seen an enormous increase in the intensity and scope of cross-border interactive relationships over the last two decades – be they economic transactions or processes of informational, cultural and political exchange (Alasuutari 2000; Castells 2004; Dreher 2006; Held et al. 2000; Robertson 1990; Sutcliffe and Glyn 1999; Raab et al. 2008). Particularly through the rapid advances in information technologies in recent years, the fall of the Iron Curtain with the resulting abrupt opening of new markets and the integration of Asian countries into the world market, there has been a marked intensification of cross-border exchange between modern states that has attained a new and previously unattained quality.

Nowadays, most social scientists assume that the globalisation process is characterised by the simultaneous coaction of four macrostructural trends that have become increasingly dominant, particularly since the 1980s (see also, Figure 3.1). These are:

1. The increasing internationalisation of markets and the associated growth in competition between countries with very different wage and productivity levels as well as different social and environmental standards (particularly since the fall of the Iron Curtain and the integration of East European and Asian nations into the global market).
2. The intensification of competition between nation states and the resulting tendency for modern states to reduce business taxes and to engage in deregulation, privatisation and liberalisation while also strengthening the market as a coordination mechanism.
3. The rapid worldwide networking of persons, companies and states through new information and communication technologies and, as a result, the increasing global interdependence of actors along with the increasing acceleration of social and economic interaction processes.
4. The fast growth in the importance of globally networked markets and the accompanying increase in the interdependence and volatility of local markets that are ever more vulnerable to scarcely predictable social, political and economic 'external shocks' and events throughout the

world (such as wars, economic crises, sub prime mortgage turbulences, oil price shocks, consumer fashions, technological innovations).

In the past years, globalisation has certainly increased productivity and improved the general standard of living in broad population strata of modern societies. Nonetheless, it is simultaneously accompanied by a growth in unexpected market trends in an increasingly changing global economy, by more rapid processes of social and economic change, by an ever stronger decline in the predictability of economic and social trends and, as a result of this, by a general increase in uncertainty (see Figure 3.1).

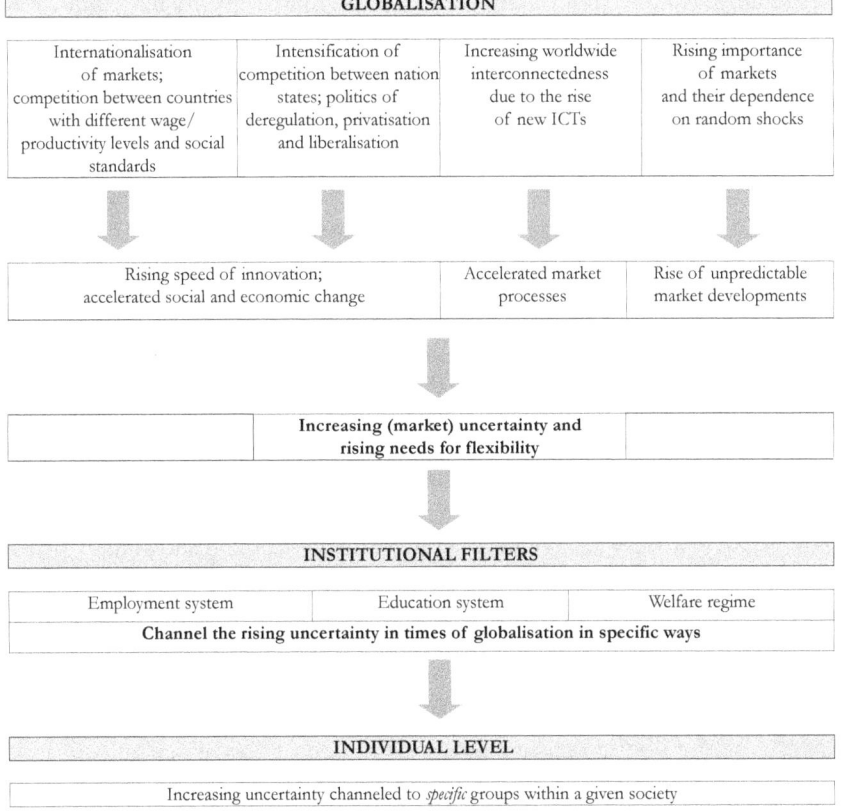

Figure 3.1   Globalisation and rising uncertainties in modern societies

Source: Own illustration following Mills and Blossfeld (2005).

As a consequence, globalisation has led to a significant shift in the power relations on the labour market. Employers increasingly try to shift their own greater market risks due to the globalisation process and their resulting needs for flexibility on to their employees, and so-called asymmetric relationships are asserting themselves on the labour market (cf. Breen 1997). Asymmetric relationships are characterised by the stronger negotiating party – in this case, the employer – retaining the option of withdrawing from the relationship should circumstances require. The other, weaker party – that is, the employee – has to accept the decisions of the stronger party without any sharing of the risk. Examples of asymmetric employment relationships are subcontracts or short-term work contracts. In both cases, employers enter into only a short-term commitment with their employees and retain the option of dismissing them again as soon as, for example, markets or orders decline.

So far, there is no consensus in sociological research with regard to how these changes in the labour market and the rise in uncertainty have influenced the development of social inequalities in modern societies. Currently, two opposite interpretations of the effects of globalisation on the development of social inequality structures can be found (Bernardi 2000; Mills et al. 2006). The first perspective – reflected in the works of Beck (1992) and Giddens (1990, 1994, 1998), for example – argues that present-day societies can no longer be characterised as 'class societies', but need to be understood as 'risk societies'. In the course of the globalisation process, new forms of risks and uncertainty have emerged and have become generalised across all social strata, thereby breaking down the logic of the traditional class structure. The second perspective suggests the opposite, namely the strengthening of existing social inequalities in the globalisation process. Breen (1997) argues that especially already disadvantaged labour market groups suffer from increasing uncertainties in the globalisation process. While well qualified employees still enjoy high protection against labour market risks, employers tend to impose asymmetric relationships especially on the less qualified and on the outsiders of the labour market. Thus, the increase of risks and uncertainty in the globalisation process is channelled systematically to the disadvantage of the lower skilled and less qualified workers, thereby amplifying the importance of social characteristics, such as education and class.

Changing Life Courses in the Globalisation Process: Results of the GLOBALIFE Project

The GLOBALIFE project is the first empirical study of the effects of the globalisation process on individual life courses, and thereby on the development of social inequality in a range of societies in Europe and North America. A sequence of four research phases was used to analyse central transitions in the life course and employment career of women and men:

1. the transition from youth to adulthood and the accompanying process of becoming established on the labour market during this period along with its effects on family formation and fertility (Blossfeld et al. 2005);
2. the transitions during the employment course of men in mid-career (Blossfeld et al. 2006);
3. the transitions during the employment course of mid-life women, paying particular attention to family formation and motherhood (Blossfeld and Hofmeister 2006); and
4. the changes in late careers and the transition to retirement (Blossfeld et al. 2006).

In each of these four single research phases, comparable country studies were set up for a total of 17 OECD member states in order to reconstruct the effects of the globalisation process on the respective life-course transitions in each national context. A detailed analysis of each country's institutional and cultural characteristics was used to group countries into five different welfare regimes:[2] liberal (Canada, Great Britain and the United States), conservative (Germany, The Netherlands and France), social-democratic (Norway, Denmark and Sweden), family-oriented (Italy, Spain, Ireland and Mexico) and post-socialist (Estonia, Hungary, the Czech Republic and Poland). Hence, the GLOBALIFE project covered an exceptionally broad spectrum of advanced countries with highly differing institutional and cultural structures channelling the effects of the globalisation process on individual life courses in very different ways. Furthermore, it has been one of the first studies that has analysed thoroughly the repercussions of the globalisation process in the post-socialist countries of the former 'Eastern Bloc'. The results of the four project phases are documented extensively in four volumes (Blossfeld et al. 2005; Blossfeld et al. 2006; Blossfeld and Hofmeister 2006; Blossfeld et al. 2006). In this chapter, we summarise the key findings of the four project phases.

---

2   This was based on Esping-Andersen's (1990) regime classification, which is well established in comparative social research and distinguishes between liberal, social-democratic and conservative welfare regimes. It was supplemented with the family-oriented regime covering not only the South-European countries, which Ferrera (1996) views as a distinct type of regime, but also Mexico and Ireland. What these family-oriented countries studied in the GLOBALIFE project have in common is that they aim to secure individual welfare by placing more trust in familial solidarity and networks than in universal state services. The post-socialist regime is a further extension to Esping-Andersen's classic welfare state typology. Although this regime is a relatively heterogeneous cluster (see below), all these countries share the abrupt transition from a comparatively isolated, planned socialist economy to an open market economy exposed to the forces of globalisation.

Youth and Young Adults: The Losers of Globalisation

The first research phase of the GLOBALIFE project took a cross-national comparative perspective to compare how adolescents and young adults manage to enter the labour market under the conditions of globalisation, and how changed labour market entry and early career patterns in young persons' lives impact on familial decisions such as entering marriage or having a child. The GLOBALIFE project analyses show that young persons face a strong increase in uncertainties when entering the labour market (Blossfeld et al. 2005; Golsch 2005; see Table 3.1) that manifest particularly in the form of a major increase in precarious, atypical forms of employment (e.g., short-term jobs, part-time jobs, precarious forms of self-employment and, compared with older cohorts, lower income). These developments tend to make young people the 'losers' of the globalisation process. At first glance, this seems to be contra-intuitive because the young generation is far more educated than older ones and many of these young people have spent a longer part of their life abroad. However, they are affected particularly strongly, because they frequently lack job experience and strong ties to business networks, particularly in internal labour markets. Often they are unable to fall back on established contacts and do not possess the negotiating power to demand stable and continuous employment. Thus, it is comparatively easy for employers and unions to adjust young people's work contracts and make them more flexible and less advantageous at their expense.

However, the concrete effects of the globalisation process on the labour market positions of young adults vary according to the nation-specific welfare-state and labour market regime (see Table 3.1). The strong insider-outsider markets of Southern Europe (but, in part, also Germany) reveal increasing phases of unemployment and/or above all short-term work contracts (Bernardi and Nazio 2005; Kurz et al. 2005; Simó Noguera et al. 2005). Particularly in Southern Europe, forms of precarious self-employment can be found. In The Netherlands, there is a massive increase in part-time jobs for young women and men (Liefbroer 2005); and in the open employment systems of the liberal countries (United States, Great Britain), the effects of the globalisation process are manifesting across generations particularly in increasing income losses for young persons (Berkowitz King 2005; Francesconi and Golsch 2005).

Independent of the national context, education is clearly becoming more and more important in the globalisation process (Blossfeld et al. 2005, see Table 3.1). Poorly qualified labour market entrants are hit particularly hard by the global changes. This is how globalisation generally reinforces the social inequalities within the young generation, because individual (human capital) resources gain in importance through the growing relevance of the market and individual competition.

The increasing experience of employment uncertainties in young adulthood has consequences for familial decision processes. Growing economic and temporal uncertainties lead young people more and more to postpone or even to forgo family formation (Blossfeld et al. 2005, see Table 3.1). On the societal level, this leads to a dilemma, because not only improved conditions for labour market flexibility in the sense of greater competitiveness but also rising birthrates are viewed as desirable.

Young adults have developed four behavioural and adaptive strategies as a reaction to growing uncertainties in the life course (Mills et al. 2005):

1. they increasingly postpone decisions requiring a long-term commitment; the youth phase becomes more and more of a 'moratorium', and transitions to gainful employment often take a chaotic course;
2. they switch increasingly to alternative roles instead of employment (e.g., they spend longer in the education system instead of letting themselves be defined as 'unemployed');
3. they are increasingly forming more flexible forms of partnership (e.g., consensual unions) that permit an adaptation to rising uncertainty without having to make long-term commitments (Nazio and Blossfeld 2003; Nazio 2008);
4. particularly in the family-oriented welfare states of Central and Southern Europe, they have developed gender-specific strategies to deal with uncertainty: men are increasingly less able to guarantee any long-term income security as the 'breadwinner' for a household, often leading to delay in family formation. In contrast, many unqualified women who 'have nothing to lose' react to the growing uncertainties on the labour market by turning to the security of the family and the traditional roles of mother and housewife (as a strategy to reduce uncertainty). Conversely, the tendency for highly qualified women to have children in increasingly uncertain labour markets depends on whether they can protect their careers by making family and career compatible. When childcare facilities are underdeveloped, as is particularly the case in Southern Europe, many qualified women decide in favour of their careers rather than for children (Bernardi and Nazio 2005; Simó Noguera et al. 2005).

Hence, a paradoxical outcome of the globalisation process is that precisely in traditional family-oriented societies, the birth rate is declining markedly because of the growing experience of employment uncertainties for young men and the incompatibility of family and career for qualified women. A similar restrained fertility behaviour can also be found in the transformation countries of Eastern Europe in which uncertainties have grown enormously since the collapse of socialism (Katus et al. 2005; Róbert and Bukodi 2005). Whereas demographic approaches attribute this change merely to a 'value shift' in modern societies (cf. e.g., Surkyn and Lesthaeghe 2002; van de Kaa 1987, 2001), a globalisation approach oriented toward increasing structural uncertainties is in a position to resolve the paradox between an often marked young people's desire for children and the failure to actually realise this desire in young adulthood. Founding a family calls for at least a minimum of economic and social certainty regarding the future that, under the global conditions of increasing labour market uncertainty, can often be achieved only after a long transition period following the exit from the education system. Therefore, economically and socially speaking, young people who forgo having children are making a rational response to structural trends. In Scandinavian countries in which the state provides comparatively generous universal services for young adults and families along with childcare facilities while engaging in an active employment policy, the birth rate is comparatively high, though still below the net reproduction rate (Bygren et al. 2005; Nilsen 2005).[3]

It is important to point out in this context that it is not the *absolute* level of uncertainty that is decisive for the structuring of decisions on family formation, but the subjectively perceived *relative* level of uncertainty in the specific country's labour force (Blossfeld et al. 2005). In each country, young adults compare themselves in daily life with 'significant others' (such as friends, relatives, acquaintances) when judging their individual labour market situation. In the United States, for example, the absolute level of uncertainty for the young generation is higher as a whole than in many European countries. People lose their jobs more frequently, but the unemployed can rely on soon finding another job, that is, becoming an 'insider' again, because of the low mobility barriers on the labour market. This is why labour market uncertainty, career mobility and flexibility possess a different social significance in the United States. Subjectively, they are perceived differently than in the insider-outsider markets of Europe in which 'being an outsider' often means an identity-threatening, long-term exclusion from work in a climate in which flexible work arrangements are generally viewed as only a stop-gap solution on the way toward a permanent job. Young persons in flexibilised forms of employment in the European insider-outsider markets therefore experience their fate as being far more negative than their peers in the United States.

---

3 Ireland is a notable exception regarding the effects of globalisation on the family foundation process. It is almost a perfect example of how a country can profit from globalisation by embracing competition on the world market, promoting free trade relations and giving tax incentives to foreign investors (Layte et al. 2005). Ireland, which achieved almost full employment at the end of the 1990s, is the only country in the GLOBALIFE project in which globalisation has led to a *decline* in employment relation uncertainties. Correspondingly, there has also been a marked upturn in marriage rates and birth rates since the mid-1990s. Nonetheless, it should be noted that Ireland's competitive advantage over other modern industrial states is due particularly to the fact that other countries have not yet cut taxes on corporation so drastically. If all countries were to follow the same path, the one country's advantage over the others would disappear. There are some developments in this direction in Eastern Europe at the time being. It remains therefore open to which extent Ireland can keep its special position.

Men in Mid-career: Are They the Winners of Globalisation?

The second phase of the GLOBALIFE project focused on the employment trajectories of mid-career men. Results showed that men from younger birth cohorts are certainly confronted with a somewhat greater labour market uncertainty than older birth cohorts (Blossfeld et al. 2006). Nonetheless, globalisation in no way leads, as frequently assumed (cf. Beck 1992), to an increasing erosion of traditional male employment relationships or to the spread of 'patchwork careers' and 'job hopping'. Quite the opposite. In contrast to the described developments for youth in their early employment lives, the careers of well-qualified male employees who are established on the market are still very stable and broadly protected from any flexibilisation by employers (see Table 3.1).

Paradoxically, although the globalisation process forces companies to respond more flexibly and therefore to try to reduce their own market risks and pass these on to their employees by imposing more short-term-oriented employment relationships, a completely flexible workforce is neither desirable nor efficient from the company perspective. It would threaten the reliable and permanent cooperation between management and qualified staff. Indeed, studies show that marked flexibilisations in companies greatly reduces their staff's willingness to cooperate, their work motivation and their company loyalty (cf., e.g., Köhler et al. 2005). In times of greater (international) competition, a secure and long-term cooperation with qualified and experienced staff is still important for employers in order to ensure the flow of information in the company, productivity and also innovations. The comprehensive introduction of flexibilised employment relationships additionally carries the threat that a company will be faced with high recruitment costs and qualification losses when hiring new employees (Breen 1997; Mills et al. 2006). To summarise, employers have no interest in withdrawing from long-term commitments to *all* the employees on their staff. Therefore, they try not to threaten the *trust relationship* to those in qualified positions. These ambivalent company goals in the globalisation process, namely, flexibilisation on the one side but stability and continuity on the other, lead to an increasing *segmentation* of labour into core groups and peripheral groups, into insiders and outsiders (Blossfeld et al. 2006). As the results of the GLOBALIFE project show, the demarcation cut-off between those belonging to the insiders and those belonging to the outsiders is being raised increasingly higher. Essentially, this is disadvantaging the lowly qualified and people who are less established on the labour market such as those completing education, young adults and women. Qualified and experienced male employees, in contrast, are broadly protected from the (labour-market) flexibilities in the globalisation process.

Even if qualified men in mid-career can generally be described as the winners of the globalisation process, there are still major differences in the various regimes studied in the GLOBALIFE project and also within the group of men (Mills and Blossfeld 2006, see Table 3.1). In all the countries of the GLOBALIFE project, it was possible to identify a quantitatively significant group of mostly low-qualified male who are increasingly long-term unemployed and who fail to successfully re-enter the labour market – in particular in the insider/outsider labour market countries. There is a further group of men in mid-career who 'oscillate' between unemployment and jobs with a low occupational status. Again, these are fairly unskilled men. Therefore, membership in the male group of 'risk employees' depends decisively on *individual resources*, in particular, on the level of education and the occupational class (Mills and Blossfeld 2006). It is particularly qualified and experienced employees in the midst of their career who enjoy comprehensive protection; they may be called the winners of globalisation. Male employees with low human capital resources, however, prove to be exceptionally vulnerable on globalised labour markets (see Table 3.1).

The relative size of this group of 'male globalisation losers' varies from country to country; nonetheless, it is comparatively large particularly in the United States and in some of the former socialist countries of Eastern Europe. In contrast, the *majority* of mid-career men in countries with conservative, family-oriented and social democratic structures enjoy a high degree of stability in their employment careers, because they are comprehensively protected by country-specific labour market and welfare state regulations (Blossfeld et al. 2006).

Women in Mid-life: Marginalisation in the Globalisation Process

The effects of the globalisation process on the middle phase of women's life courses differ markedly from those on men in mid-career. Whereas, as shown above, mid-career male employment courses prove to remain comparatively stable in the globalisation process, women reveal a trend toward *marginalisation* on the labour market (Blossfeld and Hofmeister 2006; see Table 3.1). As the GLOBALIFE study has shown, although women have become increasingly integrated into modern labour markets as a result of the globalisation process in a series of countries, this integration has often been quite precarious.

Several parallel developments since the beginning of the 1960s have led middle-aged women to become increasingly active on the labour markets in Europe and North America. Due to quantitative and qualitative improvements in their access to education, young women show not only a greater interest in their own careers but also have improved their preconditions for successful labour market participation compared with women in earlier generations. Moreover, the increasing instability of families along with the declining security in the employment careers of husbands in more recent generations in some countries have contributed to greater female participation on the labour market. As a result of these changes, women's incomes have become increasingly important for the material security of women and their families.

Table 3.1 Life courses in the globalisation process

	Young adults	Mid-career men	Mid-life women	Late-career employees
Main effect	Increased employment uncertainty, resulting in postponed family formation	Relatively persistent levels of employment stability	Marginalisation on the labour market	Increased risks of forced employment withdrawal
Regime-specific effect	*Southern European and conservative countries*: Marginalisation of youth as labour market outsiders due to increasing affectedness by precarious employment; very strong impact on family formation and childbirth *Post-socialist countries*: Even stronger employment insecurities; dramatic effects on family formation *Social-democratic countries*: Relative shielding of youth and family from employment uncertainty *Liberal countries*: Employment uncertainty counterbalanced by open labour market structures; relatively low impact on family formation due to modest changes in subjectively perceived uncertainty	*Conservative, Southern European and social-democratic countries*: Constantly high level of stability *Post-socialist and liberal countries*: Modest increases in employment flexibility among mid-career men	*Conservative and Southern European countries*: Increasing integration of women into employment, but only as secondary earners in less stable employment *Social-democratic countries*: Relative stability of employment levels due to active state supports *Liberal countries*: Increasing need to support family income pushes women into (flexible) employment *Post-socialist countries*: Loss of full employment status after the fall of the Iron Curtain	*Conservative and Southern European countries*: Highest rate of early exit, largely buffered by generous pension systems *Social-democratic countries*: Late career exits and high employment stability fostered by means of active labour market policies *Liberal countries*: Late career exits, but relatively high employment mobility *Post-socialist countries*: Implementation of differential strategies (Hungary and Czech Republic = conservative strategy; Estonia = liberal strategy)
Individual-level effects	Increasing importance of education as a key factor to become established in the labour market	Losses in employment stability largely concentrate on low(est) qualified men and low occupational classes	Increasing importance of employment experience and course transitions; comparatively higher importance of educational attainment	Overall inter-individual variation less pronounced than for other life course transitions; comparatively higher importance of human capital in liberal countries

Source: Own illustration.

This increasing work *supply* of women on the one side has been matched by a growth in the *demand* for female labour on the other side in the expanding (private and public) service sector brought about by the globalisation process (Blossfeld and Hakim 1997).

Despite their growing integration into the labour force, it is almost exclusively women who continue to perform the unpaid familial and care duties in all modern societies. During the family phase, married couples tend far more to invest in the continuing working career of the husband rather than that of the wife (Blossfeld and Drobnič 2001). This practice not only limits women's earning capacities but can also impair their continuity of employment and career chances in the long term, particularly when wives give up their jobs completely in favour of those of their husbands, or adapt them to those of their husbands in terms of time or space. Disadvantages in their employment careers are not just experienced by those women who *actually* interrupt their employment careers for familial reasons. Even those not planning such a break are frequently considered to be possibly or probably planning to do it – and this argument is used to deny them jobs, promotions and further training opportunities solely because of their gender (the so-called 'statistical discrimination') (Blossfeld and Hakim 1997). As a result, women are overrepresented in the flexible forms of work emerging within the globalisation process, such as part-time work (Blossfeld and Hofmeister 2006). Employers legitimise this concentration of flexible forms of work on women by pointing repeatedly to their deficits in work experience compared with men and the greater probability of a later employment interruption. In contrast to those forms of employment that, although flexibly organised, are basically secure, the flexibility desired by companies is frequently tied to insecure, precarious forms of employment that grant employers short-term decision scope (Blossfeld and Hakim 1997). Jobs with such goals frequently do not correspond to the *personal* flexibility needs of women, that is, the option of being able to interrupt, reduce, or plan work flexibly in order to fulfil simultaneous care duties. However, women often accept such flexible and less secure forms of work so that they can meet their familial obligations when other measures to promote the compatibility of career and family are lacking or insufficient.

The results of the GLOBALIFE project have shown that various forms of disadvantage on the labour market are accumulating increasingly on women in the globalisation process: precarious, insecure and low-paid jobs; part-time jobs or jobs with variable work shifts; jobs with little autonomy, control, or responsibility; and jobs with only slight possibilities of promotion and high risks of downward career mobility or unemployment (Hofmeister and Blossfeld 2006). Hence, globalisation contributes across all countries to a marginalisation of women as 'outsiders' of the labour market.

Despite this general tendency toward a marginalisation of women on the labour market, the GLOBALIFE project once again reveals different patterns of development in the various regimes (see Table 3.1): In conservative and, to a lesser extent, Southern European countries, the creation of new, flexible job options in the globalisation process has contributed to a better *integration* of women into national labour markets, even though on a precarious basis (Buchholz and Grunow 2006; Kalmijn and Luijkx 2006; Pisati and Schizzerotto 2006; Simó Noguera 2006). An increasing integration of women can also be observed in the liberal states that offer little state support for families and pursue a 'laissez-faire' labour market policy. However, in contrast to the conservative and Southern European countries, it is above all low-qualified women who enter the labour market to support their families financially, and this is frequently in low-income jobs (Hofmeister 2006). The social-democratic and also former socialist countries, which already possess a long tradition of high female employment rates, reveal diverging trends in the globalisation process: Sweden has at least managed to stabilise its exceptionally high proportion of employed women while retaining job security in the globalisation process (Korpi and Stern 2006). In contrast, the proportion of working women in Denmark has dropped since the introduction of vacation regulations (Grunow 2006; Grunow and Leth-Sørensen 2006). In the former socialist countries, in contrast, integration into the world economy has led to stagnating or even negative developments in female employment rates (Bukodi and Róbert 2006; Hamplová 2006; Helemäe and Saar 2006; Plomien 2006).

As with men, individual resources, particularly in the form of education qualification, are of central significance for the course of employment careers in middle-aged women (Hofmeister and Blossfeld 2006; see Table 3.1). This education effect is also cumulative: well-educated young women in modern, knowledge-based business fields possess the greatest chances of avoiding unemployment and advancing their careers. In contrast, the disadvantaged in the globalisation process are women with few qualifications, little work experience and earlier childrearing breaks in employment along with those who have more frequent and longer phases of unemployment. They face a particularly high risk of having to work in insecure and precarious jobs or of becoming (repeatedly) unemployed (Blossfeld and Hofmeister 2006).

Late-career Employees: Increasingly Confronted with Accelerating Structural Change in the Globalisation Process

The fourth and final phase of research in the GLOBALIFE project showed that the work careers of older employees have also undergone a remarkable transformation in the globalisation process. The cause of this change in later work careers has been an increasing discrepancy between the growing demands for employment flexibility and the limited possibilities for flexibilising the work conditions and qualification profiles of older employees that are becoming even further reinforced through globalisation (Buchholz et al. 2006).

As mentioned above, companies facing global market competition have a growing need to be able to continuously adapt flexibly to changing economic conditions. However, in light of these new flexibility demands through globalisation, older employees reveal several *competitive disadvantages* compared with their younger competitors (Buchholz et al. 2006). They mostly possess only obsolescent technological knowledge and vocational qualifications that make it harder to adapt them to the greater technological change in the globalisation process. As a consequence, the decisive advantage that older employees used to have, namely, their work *experience* compared with young labour market entrants, is rapidly losing importance. At the same time, the costs of requalifying older employees through further training programmes and on-the-job training measures are often too high for employers because of the poor returns to investments due to the few years of employment left. Young employees, in contrast, possess more modern qualifications, their training is often to the latest occupational and technological standards and the costs of any necessary further training can be recouped over a longer period of time. In a number of countries, older employees also earn higher 'seniority wages' through the logic of internal labour markets. This often makes them far more expensive than their younger competitors, without this difference in pay being linked from the employer's perspective with higher productivity. In addition, their jobs often enjoy strong employment protection that is very hard for companies to overcome. In contrast, young employees have been exposed to a destabilisation and flexibilisation in their employment careers within the globalisation process (Blossfeld et al. 2005). Their work contracts are correspondingly less regulated and they often earn less.

In summary, companies perceive older employees as being less flexible, inadequately qualified and cost-intensive in the globalisation process. Hence, it is in the interest of not only companies but also policy makers concerned with the attractiveness of their national business location to find solutions for this discrepancy between increasing demands for flexibility and the limited flexibilisation potential of older employees (Blossfeld et al. 2006).

One option for resolving this contradiction is to offer attractive financial incentives for early employment exit. Indeed, comparative labour market data reveal a trend toward an increasingly early retirement of older employees in almost all Western industrial societies since the 1970s (cf. OECD 2006). Detailed analyses in the GLOBALIFE project have shown, however, that the *magnitude* of this early retirement trend varies markedly between different OECD countries (cf. Hofäcker and Pollnerová 2006; Hofäcker et al. 2006, see Table 3.1). To cope with the problematic labour market situation of older employees in the globalisation process, modern societies therefore appear to pursue various strategies that can be classified according to three different ideal types (cf. Buchholz et al. 2006):

Central and Southern European states in particular pursue a strategy of one-sidedly promoting the *employment exit* of older employees in order to cope with the global competitive pressure and the structural changes in the economy (Blossfeld et al. 2006). The discrepancy between growing demands for flexibility in the globalisation process and the flexibilisation potential of older employees is particularly large in these countries. Due to a highly standardised education system which largely restricts vocational training to the early life course and which offers few possibilities of in-service or further training, older employees often have major qualification disadvantages compared with their younger labour market competitors. At the same time, extensive protection from dismissal and a well-established system of seniority wages limit the possibilities of making their work contracts more flexible. As a way out of this dilemma, the Central and Southern European states have extended existing early retirement options and created new early retirement pathways in part by additional interim welfare state provisions to bridge the gap to full retirement (Beckstette et al. 2006; Buchholz 2006, 2008; Henkens and Kalmijn 2006).

This 'early retirement strategy' has advantages for all concerned on the labour market. Older employees who retire early are ensured an adequate standard of living through high (early) pensions; companies gain a 'socially acceptable' opportunity to implement rationalisations and restructurings; governments profit from such strategies by raising the attractiveness of the national production location for companies while simultaneously relieving pressure on the national labour market. Nonetheless, despite these clear advantages, applying the early retirement strategy to solve the flexibilisation dilemma is a cost-intensive option for modern societies. As the results of the GLOBALIFE study show, other societies have broadly rejected such early retirement measures and by promoting lifelong learning and an active employment policy, they have made it possible for older employees to adapt flexibly to the challenges of structural and technological change. Empirically, two strategies for the *employment maintenance* of older employees can be distinguished (Blossfeld et al. 2006).

Liberal states (Great Britain, United States) broadly follow a model of maintaining older employees through *market mechanisms*. The policy for adjusting older workers to new flexibility demands is to place broad trust in a flexible labour market and an only marginally standardised education and training system. Low mobility barriers on the labour market and a decentralised organisation for acquiring relevant qualifications 'on-the-job' enable older employees to adapt flexibly to changing demands through labour market mobility. At the same time, low state pensions and a strong emphasis on private schemes based on capital investments or company pensions limit the possibilities of an early exit from employment. Older employees in the liberal states have correspondingly long employment careers and often retire comparatively late. Because of the far-reaching non-involvement of the state and the trust in market mechanisms, the liberal system nonetheless tends to strengthen social inequalities on the labour market beyond retirement age.

Employees with meager financial resources sometimes still have to carry on working after retirement age, or they return to the labour market because they are unable to survive on their pensions alone (cf. Warner and Hofmeister 2006).

In contrast, the social-democratic states in Scandinavia (Sweden, Norway and Denmark) actively engage in supporting the ability of older employees to adapt to the flexibility demands due to globalisation (i.e., by maintaining older employees through *state mechanisms*). An active labour market policy as well as state promotion of life-long learning and further vocational qualification help to keep them employable so that their employment careers are more continuous and stable than in the liberal model oriented toward labour market mobility. At the same time, pension systems with lower incentives for early retirement favour a long work career. Although the social-democratic states have also introduced some isolated early retirement options in response to growing unemployment rates, the labour market participation of older employees remains markedly higher than the international average (cf. Aakvik et al. 2006; Hofäcker and Leth-Sørensen 2006; Sjögren Lindquist 2006).

Summary

The spatial dynamics caused by the globalisation process – for example, the rapid worldwide networking through new information and communication technologies and the increasing internationalisation of markets – have strongly affected the mobility of labour and capital and, as a result, the (re-)structuring of social inequalities. This chapter summarised the results of the GLOBALIFE project carried out from 1999 to 2006 at the Universities of Bielefeld and Bamberg and funded by the Volkswagen Foundation. The goal of the project was to perform an empirically based cross-national comparative analysis of the effects of globalisation on labour market flexibilisation in different countries that each have their own specific institutional contexts. This social scientific project is the first study which systematically examined the effects of the globalisation process on individual life courses and employment careers and thereby the structures of social inequality in modern societies from an international comparative perspective.

The research focused on four central phases of the life course:

1. the transition from youth to adulthood and the establishment on the labour market that occurs during this phase along with its effects on family formation and fertility;
2. the employment course of men in mid-career;
3. the employment course of women paying particular attention to family development and motherhood; as well as
4. the late employment career and the transition to retirement.

As the results of the GLOBALIFE project show, the globalisation process has impacted very differently on these different phases of the life course (see Table 3.1): qualified men in mid-career are broadly protected from the effects of globalisation. Globalisation has in no way led to the frequently assumed increase in the erosion of traditional male employment relationships or to a massive spread of 'patchwork careers'. In contrast, the employment careers of well-qualified (male) persons who have established themselves on the labour market continue to be very stable in modern societies. In contrast, young adults, women in mid-life and people approaching retirement have had to accept a clear change in their life courses as a result of the globalisation process, although the ways in which the life courses of these groups have changed vary greatly. In all, it can be seen that it is particularly young adults who can be described as the losers of the globalisation process. Their labour market situation has deteriorated profoundly in recent years. This group sees itself as being confronted with fundamental uncertainties in working life, particularly because it is burdened to a very great extent with flexible and precarious employment conditions without receiving any compensation for this risk.

However, not only the current phase of life strongly impacts the extent to which individuals are affected by globalisation. In addition, educational and class characteristics determine in how far an individual has to face increasing labour market risks. Under globalisation, these effects have become even stronger (see Table 3.1). Thus, the results of the GLOBALIFE project support the argument that globalisation triggers a *strengthening* of existing social inequality structures as hypothesised by Breen (1997) rather than the emergence of risk societies as proposed by individualisation theorists (see, for example, Beck 1992). Present-day societies hence still can be characterised as class societies.

The third central finding of the GLOBALIFE project is that the globalisation process has *not* led to the same outcome in the various modern societies (see Table 3.1). Hence, the results of the GLOBALIFE project explicitly contradict the frequent assumption in globalisation research of a decline in the significance of national state regulations in the course of the globalisation process leading to the *same* outcomes in different countries (Beck 2000; Meyer et al. 1992; Ohmae 1991; Treiman and Yip 1989). It is far more the case that the globalisation process in different national contexts runs up against various firmly anchored institutional structures such as welfare state institutions, specific ways of regulating labour markets, or local norms and values. These national institutions filter the increasing uncertainty due to globalisation in a *specific* way, thus leading to special forms of labour market flexibilisation that, in turn, have shaped and changed the life courses and structures of social inequality in modern societies in very different ways.

References

Aakvik, A., Dahl, S.A. and Vaage, K. (2006), 'Late Careers and Career Exits in Norway', in Blossfeld, H.-P., Buchholz, S. and Hofäcker, D. (eds), *Globalization, Uncertainty and Men's Careers in International Comparison* (Cheltenham: Edward Elgar).

Alasuutari, P. (2000), 'Globalisation and the Nation-State: An Appraisal of the Discussion', *Acta Sociologica* 43:3, 259–69.

Beck, U. (1992), *Risk Society* (London: Sage).

Beck, U. (2000), 'What is Globalisation?', in Held, D. and McGrew, A. (eds), *The Global Transformations Reader: An Introduction to the Globalization Debate* (Malden, MA: Polity Press).

Beck, U., Giddens, A. and Lash, S. (eds) (1994), *Reflexive Modernization* (Cambridge: Polity Press).

Beckstette, W., Lucchini M. and Schizzerotto A. (2006), 'Men's Late Careers and Career Exits in Italy', in Blossfeld, H.-P., Buchholz, S. and Hofäcker, D. (eds), *Globalization, Uncertainty and Men's Careers in International Comparison* (Cheltenham: Edward Elgar).

Berkowitz King, R. (2005), 'The Case of American Women. Globalisation and the Transition to Adulthood in an Individualistic Regime', in Blossfeld, H.-P., Klijzing, E., Mills, M. and Kurz, K. (eds), *Globalisation, Uncertainty and Youth in Society* (London: Routledge).

Bernardi, F. (2000), 'Globalisation, Recommodification and Social Inequality. Changing Patterns of early Careers in Italy', GLOBALIFE Working Paper No. 7 (Bamberg: University of Bielefeld).

Bernardi, F. and Nazio, T. (2005), 'Globalisation and the Transition to Adulthood in Italy', in Blossfeld, H.-P., Klijzing, E., Mills, M. and Kurz, K. (eds), *Globalisation, Uncertainty and Youth in Society* (London: Routledge).

Blossfeld, H.-P. and Hakim, C. (eds) (1997), *Between Equalization and Marginalisation. Women Working Part-Time in Europe and the United States of America* (New York: Oxford University Press).

Blossfeld, H.-P. and Drobnič, S. (eds) (2001), *Careers of Couples in Contemporary Societies. From Male Breadwinner to Dual-Earner Families* (Oxford: Oxford University Press).

Blossfeld, H.-P., Klijzing, E., Mills, M. and Kurz, K. (eds) (2005), *Globalisation, Uncertainty and Youth in Society* (London: Routledge).

Blossfeld, H.-P. and Hofmeister, H. (eds) (2006), *Globalisation, Uncertainty and Women's Careers in International Comparison* (Cheltenham: Edward Elgar).

Blossfeld, H.-P., Buchholz, S. and Hofäcker, D. (eds) (2006), *Globalisation, Uncertainty and Late Careers in Society* (London: Routledge).

Blossfeld, H.-P., Klijzing, E., Mills, M. and Kurz, K. (eds) (2006), *Globalisation, Uncertainty and Men's Careers in International Comparison* (Cheltenham: Edward Elgar).
Breen, R. (2005), 'Explaining Cross-national Variation in Youth Unemployment', *European Sociological Review* 21:2, 125–34.
Buchholz, S. (2006), 'Men's Late Careers and Career Exits in West Germany', in Blossfeld, H.-P., Buchholz, S. and Hofäcker, D. (eds), *Globalisation, Uncertainty and Late Careers in Society* (London: Routledge).
Buchholz, S. and Grunow, D. (2006), 'Women's Employment in West Germany', in Blossfeld H.-P. and Hofmeister, H. (eds), *Globalisation, Uncertainty and Women's Careers in International Comparison* (Cheltenham: Edward Elgar).
Buchholz, S., Hofäcker, D. and Blossfeld H.-P. (2006), 'Globalisation, Accelerating Economic Change and Late Careers. A Theoretical Framework', in Blossfeld, H.-P., Buchholz, S. and Hofäcker, D. (eds), *Globalisation, Uncertainty and Late Careers in Society* (London: Routledge).
Buchholz, S. (2008), *Die Flexibilisierung des Erwerbsverlaufs. Eine Analyse von Einstiegs- und Ausstiegsprozessen in Ost- und Westdeutschland* (Wiesbaden: VS-Verlag).
Bukodi, E. and Róbert, P. (2006), 'Women's Career Mobility in Hungary', in Blossfeld H.-P. and Hofmeister, H. (eds), *Globalisation, Uncertainty and Women's Careers in International Comparison* (Cheltenham: Edward Elgar).
Bulatao, R.A. and Casterline, J.B. (eds) (2001), *Global Fertility Transition. Supplement to Population and Development Review 27* (New York: Population Council).
Bygren, M., Duvander A.-Z. and Hultin M. (2005), 'Elements of Uncertainty in Life Courses. Transitions to Adulthood in Sweden', in Blossfeld, H.-P., Klijzing, E., Mills, M. and Kurz, K. (eds), *Globalisation, Uncertainty and Youth in Society* (London: Routledge).
Castells, M. (2004), *Der Aufstieg der Netzwerkgesellschaft. Das Informationszeitalter* (Opladen: Leske + Budrich).
Dreher, A. (2006), *KOF Index of Globalisation* (Zürich: Konjunkturforschungsstelle ETH Zürich).
Esping-Andersen, G. (1990), *The Three Worlds of Welfare Capitalism* (Cambridge: Polity Press).
Ferrera, M. (1996), 'The 'Southern Model' of Welfare in Social Europe', *Journal of European Social Policy* 6:1, 17–37.
Francesconi, M. and Golsch, K. (2005), 'The Process of Globalisation and Transitions to Adulthood in Britain', in Blossfeld, H.-P., Klijzing, E., Mills, M. and Kurz, K. (eds), *Globalisation, Uncertainty and Youth in Society* (London: Routledge).

Giddens, A. (1990), *The Consequences of Modernity* (Cambridge: Polity Press).
Giddens, A. (1994), 'Living in a Post-Traditional Society', in Beck, U., Giddens, A. and Lash, S. (eds), *Reflexive Modernization* (Cambridge: Polity Press).
Giddens, A. (1998), *The Third Way* (Cambridge: Polity Press).
Golsch, K. (2005), *The Impact of Labour Market Insecurity on the Work and Family Life of Men and Women. A Comparison of Germany, Great Britain and Spain* (Frankfurt am Main: Lang).
Grunow, D. (2006), *Convergence, Persistence and Diversity in Male and Female Careers – Does Context Matter in an Era of Globalisation. A Comparison of Gendered Employment Mobility Patterns in West Germany and Denmark* (Opladen: Barbara Budrich Publishers).
Grunow, D. and Leth-Sørensen, S. (2006), 'Danish Women's Unemployment, Job Mobility and Non-employment, 1980s and 1990s: Marked by Globalisation?', in Blossfeld, H.-P. and Hofmeister, H. (eds), *Globalisation, Uncertainty and Women's Careers in International Comparison* (Cheltenham: Edward Elgar).
Haferkamp, H. and Smelser, N.J. (eds) (1992), *Social Change and Modernity* (Berkeley, CA: University of California Press).
Hamplová, D. (2006), 'Women and the Labour Market in the Czech Republic: Transition from a Socialist to a Social-Democratic Regime?', in Blossfeld, H.-P. and Hofmeister, H. (eds), *Globalisation, Uncertainty and Women's Careers in International Comparison* (Cheltenham: Edward Elgar).
Held, D. and McGrew, A. (eds) (2003), *The Global Transformations Reader. An Introduction to the Globalisation Debate* (Cambridge: Polity Press).
Held, D., McGrew, A., Goldblatt, D. and Perraton, J. (2000), 'Rethinking Globalisation', in Held, D. and McGrew, A. (eds), *The Global Transformations Reader. An Introduction to the Globalisation Debate* (Cambridge: Polity Press).
Helemäe, J. and Saar, E. (2006), 'Women's Employment in Estonia', in Blossfeld H.-P. and Hofmeister, H. (eds), *Globalisation, Uncertainty and Women's Careers in International Comparison* (Cheltenham: Edward Elgar).
Henkens, K. and Kalmijn, M. (2006), 'Labour Market Exits of Older Men in the Netherlands. An Analysis of Survey Data 1979–99', in Blossfeld, H.-P. et al. (eds).
Hofäcker, D., Blossfeld, H.-P. and Buchholz, S. (2006), 'Late Careers in a Globalizing World. A Comparison of Changes in Twelve Modern Societies', in Blossfeld, H.-P., Buchholz, S. and Hofäcker, D. (eds), *Globalisation, Uncertainty and Late Careers in Society* (London: Routledge).
Hofäcker, D. and Pollnerová, S. (2006), 'Late Careers and Career Exits. An International Comparison of Trends and Institutional Background Patterns', in Blossfeld, H.-P. (eds).
Hofäcker, D. and Leth-Sørensen, S. (2006), 'Late Careers and Career Exits of Older Danish Workers', in Blossfeld, H.-P. et al. (eds).

Hofmeister, H. (2006), 'Women's Employment Transitions and Mobility in the United States: 1968 to 1991', in Blossfeld H.-P. and Hofmeister, H. (eds), *Globalisation, Uncertainty and Women's Careers in International Comparison* (Cheltenham: Edward Elgar).

Hofmeister, H. and Blossfeld, H.-P. (2006), 'Women's Careers in an Era of Uncertainty: Conclusions from a 13-Country International Comparison', in Blossfeld H.-P. and Hofmeister, H. (eds), *Globalisation, Uncertainty and Women's Careers in International Comparison* (Cheltenham: Edward Elgar).

Kalmijn, M. and Luijkx, R. (2006), 'Changes in Women's Employment and Occupational Mobility in the Netherlands: 1995 to 2000', in Blossfeld H.-P. and Hofmeister, H. (eds), *Globalisation, Uncertainty and Women's Careers in International Comparison* (Cheltenham: Edward Elgar).

Katus, K., Puur, A. and Sakkeus, L. (2005), 'Transition to Adulthood in Estonia. Evidence from FFS', in Blossfeld, H.-P., Klijzing, E., Mills, M. and Kurz, K. (eds), *Globalisation, Uncertainty and Youth in Society* (London: Routledge).

Köhler, C., Struck, O., Krause, A., Sohr, T. and Pfeifer, C. (2005), 'Schutzzone Organisation – Risikozone Markt? Entlassungen, Gerechtigkeitsbewertung und Handlungsfolgen', in Kronauer, M. and Linne, G. (eds), *Flexicurity. Die Suche nach Sicherheit in der Flexibilität* (Berlin, Sigma Verlag), 295–316.

Kohn, M.L. (ed.) (1989), *Cross-National Research in Sociology* (Newbury Park, CA: Sage).

Korpi, T. and Stern, C. (2006), 'Globalisation, Deindustrialization and the Labour Market Experiences of Swedish Women, 1950 to 2000', in Blossfeld H.-P. and Hofmeister, H. (eds), *Globalisation, Uncertainty and Women's Careers in International Comparison* (Cheltenham: Edward Elgar).

Kronauer, M. and Linne, G. (eds) (2005), *Flexicurity. Die Suche nach Sicherheit in der Flexibilität* (Berlin: Hans Böckler Stiftung).

Kurz, K., Steinhage, N. and Golsch, K. (2005), 'Case Study Germany: Global Competition, Uncertainty and the Transition to Adulthood', in Blossfeld, H.-P., Klijzing, E., Mills, M. and Kurz, K. (eds), *Globalisation, Uncertainty and Youth in Society* (London: Routledge).

Layte, R., O'Connell, P.J., Fahey, T. and McCoy, S. (2005), 'Ireland and Economic Globalisation. The Experiences of a Small Open Economy', in Blossfeld, H.-P., Klijzing, E., Mills, M. and Kurz, K. (eds), *Globalisation, Uncertainty and Youth in Society* (London: Routledge).

Liefbroer, A.C. (2005), 'Transition from Youth to Adulthood in the Netherlands', in Blossfeld, H.-P., Klijzing, E., Mills, M. and Kurz, K. (eds), *Globalisation, Uncertainty and Youth in Society* (London: Routledge).

Meyer, J.W., Ramirez, F. and Soysal, Y. (1992), 'World Expansion of Mass Education, 1970–1980', *Sociology of Education* 65:2, 128–49.

Mills, M. and Blossfeld, H.-P. (2005), 'Globalisation, Uncertainty and the Early Life-Course. A Theoretical Framework', in Blossfeld, H.-P., Klijzing,

E., Mills, M. and Kurz, K. (eds), *Globalisation, Uncertainty and Youth in Society* (London: Routledge).

Mills, M. and Blossfeld, H.-P. (2006), 'Globalisation, Patchwork Careers and the Individualization of Inequality? A 12-Country Comparison of Men's Mid-Career Job Mobility', in Blossfeld, H.-P., Klijzing, E., Mills, M. and Kurz, K. (eds), *Globalisation, Uncertainty and Men's Careers in International Comparison* (Cheltenham: Edward Elgar).

Mills, M., Blossfeld, H.-P. and Bernardi, F. (2006), 'Globalisation, Uncertainty and Men's Employment Careers. A Theoretical Framework', in Blossfeld, H.-P., Klijzing, E., Mills, M. and Kurz, K. (eds), *Globalisation, Uncertainty and Men's Careers in International Comparison* (Cheltenham: Edward Elgar).

Nazio, T. (2008), *Cohabitation, Family and Society* (New York and Abingdon: Routledge).

Nazio, T. and Blossfeld, H.-P. (2003), 'The Diffusion of Cohabitation among Young Women in West Germany, East Germany, and Italy', *European Journal of Population*, 19:1, 47–82.

Nilsen, O.A. (2005), 'Transition to Adulthood in Norway', in Blossfeld, H.-P., Klijzing, E., Mills, M. and Kurz, K. (eds), *Globalisation, Uncertainty and Youth in Society* (London: Routledge).

OECD (2006), *Ageing and Employment Policies: Live Longer – Work Longer* (Paris: OECD).

Ohmae, K. (1990), *The Borderless World* (New York: Harper Business).

Panic, M. (2003), *Globalisation and National Economic Welfare* (Basingstoke: Palgrave Macmillan).

Pisati, M. and Schizzerotto, A. (2006), 'Mid-Career Women in Contemporary Italy: Economic and Institutional Changes', in Blossfeld H.-P. and Hofmeister, H. (eds), *Globalisation, Uncertainty and Women's Careers in International Comparison* (Cheltenham: Edward Elgar).

Plomien, A. (2006), 'Women and the Labour Market in Poland: From Socialism to Capitalism', in Blossfeld H.-P. and Hofmeister, H. (eds), *Globalisation, Uncertainty and Women's Careers in International Comparison* (Cheltenham: Edward Elgar).

Róbert, P. and Bukodi, E. (2005), 'The Effects of the Globalisation Process in the Transition to Adulthood in Hungary', in Blossfeld, H.-P., Klijzing, E., Mills, M. and Kurz, K. (eds), *Globalisation, Uncertainty and Youth in Society* (London: Routledge).

Robertson, R. (1990), 'Mapping the Global Condition: Globalisation as the Central Concept', *Theory, Culture and Society* 7:1, 15–30.

Robertson, R. (1992), 'Globality, Global Culture, and Images of World Order', in Haferkamp, H. and Smelser, N.J. (eds), *Social Change and Modernity* (Berkeley, CA: University of California Press).

Simó Noguera, C. (2006), 'Hard Choices: Can Spanish Women Reconcile Job and Family?', in Blossfeld H.-P. and Hofmeister, H. (eds), *Globalisation,*

Uncertainty and Women's Careers in International Comparison (Cheltenham: Edward Elgar).

Simó Noguera, C., Castro, T. and Bonmatí A.S. (2005), 'The Spanish Case. The Effects of the Globalisation Process on the Transition to Adulthood', in Blossfeld, H.-P., Klijzing, E., Mills, M. and Kurz, K. (eds), *Globalisation, Uncertainty and Youth in Society* (London: Routledge).

Sjögren Lindquist, G. (2006), 'Late Careers and Career Exits in Sweden', in Blossfeld, H.-P., Buchholz, S. and Hofäcker, D. (eds), *Globalisation, Uncertainty and Late Careers in Society* (London: Routledge).

Surkyn, J. and Lesthaeghe, R. (2002), *Value Orientations and the Second Demographic Transition (SDT) in Northern, Western, and Southern Europe: An Update* (Brussels: Vrije Universiteit Brussel).

Sutcliffe, B. and Glyn, A. (1999), 'Still Underwhelmed: Indicators of Globalisation and their Misinterpretation', *Review of Radical Political Economics* 31:1, 111–32.

Treiman, D.J. and Yip, K.-B. (1989), 'Educational and Occupational Attainment in 21 Countries', in Kohn, M.L. (ed.), *Cross-National Research in Sociology* (Newbury Park, CA: Sage).

Van de Kaa, D.J. (1987), 'Europe's Second Demographic Transition', *Population Bulletin* 42:1, 48–57.

Van de Kaa, D.J. (2001), 'Postmodern Fertility Preferences: From Changing Value Orientation to New Behaviour', in Bulatao, R. and Casterline, J. (eds), *Global Fertility Transition*, supplement to *Population and Development Review* 27 (New York: Population Council), 17–52.

Warner, D. and Hofmeister, H. (2006), 'Late Career Transitions among Men and Women in the United States', in Blossfeld, H.-P., Buchholz, S. and Hofäcker, D. (eds), *Globalisation, Uncertainty and Late Careers in Society* (London: Routledge).

This page has been left blank intentionally

Chapter 4
Metaphors of Mobility – Inequality on the Move

Jonas Larsen and Michael Hviid Jacobsen

Introduction

Spatial mobility has become a key feature of global modernity and social theory within the last few decades. Indeed, major sociologists like Zygmunt Bauman and John Urry – as part of a 'new mobility paradigm' – argue that 'nowadays we are all on the move' (Bauman 1998a, 77) and 'sometimes it seems as if the world is all on the move' (Urry 2007, 3). And yet it is also increasingly evident that the access to and benefits of these mobilities and 'time-space compressions' are unequally distributed both within national societies and across the globe (see Chapter 1). People travel under different circumstances and for different reasons, and so we should differentiate between various forms of physical travel and understand how they are caught in various power geometries of everyday life. (Massey 1991, 317). Today, as Zygmunt Bauman describes, access to the 'right' sort of travel, to the 'right' kind of mobility, becomes a major stratifying factor: 'Mobility climbs to the rank of the uppermost among the coveted values – and the freedom to move, perpetually a scarce and unequally distributed commodity, fast becomes the main stratifying factor of our late-modern or postmodern times' (Bauman 1998a, 2). These inequalities are largely enhanced by the 'digital divide'. Even though internet access and wireless communication diffuse around every corner of the world with unprecedented speed, there are still profound inequalities of what John Urry (2007) calls 'virtual travel' and 'communicative travel'.

This chapter does not address such mobile inequalities empirically but metaphorically, and we do so by making the first analysis and comparison of Zygmunt Bauman and John Urry's recent mobile and spatial metaphors though a lens of social inequality (but see Jokinen and Veijola's (1997) feminist critique of the 'maleness' of Bauman's metaphors). The work of Bauman and Urry permeates with such metaphors and we aim to highlight how their playful yet useful utilisation of metaphors can be productive in illuminating new understandings and concepts of mobile and spatial inequalities that may guide contemporary and future research and theories. We begin with a theoretical section on metaphors asserting how metaphors are specifically fertile tools with which to both familiarise and defamiliarise the social world as we know it.

We show how the thought processes involved in a metaphorical redescription of social life – transformation, transference, transgression, transportation, transmutation and transcendence – potentially point beyond existing social reality and sociological understandings while, at the same time, allowing for a more creative and enhanced perspective on the real world. Then we move on to discuss Bauman's famous metaphors of 'the tourist' and 'the vagabond' as well as the less well-known ones of 'escape', 'exit' and 'mismeeting'. With regard to Urry we argue that his recent metaphors of 'network capital' and 'meetingness' are important in relation to understanding the nature and significance of social inequality in mobile networked societies. Following this, a conclusion critically assessing the 'productiveness' of Bauman's and Urry's metaphors, and the piece is concluded by arguing for the need for developing new metaphors that break with the somewhat dualistic or binary thinking between the 'mobility rich' and 'mobility poor' that permeates especially Bauman's writing.

The Magic of Metaphors

In *Sociology as an Art Form*, Robert A. Nisbet contended how 'human thought in the large is almost inconceivable apart from the use in some degree of metaphor' (2002 [1976], 33). Similarly, in *Metaphors We Live By*, George Lakoff and Mark Johnson (1980) asserted how most human thought processes are metaphorical and how humans therefore navigate in life by way of such metaphors. However, we not only 'live' by metaphors, we also 'research' by them – the way we comprehend, describe, analyse and diagnose social reality is, in fact, overwhelmingly metaphorical. Although an apparently integral part of everyday language as well as scientific reasoning, there has been no shortage in philosophical and social scientific discourse of contemptuous and critical attitudes towards metaphors. Metaphors are, it is claimed, not a natural part of the realm of science but instead belong to the realm of poetry and literature (see Lakoff and Turner 1989). Therefore, metaphors have for decades been attacked for obscuring the vision and obfuscating the language of science. However, by describing what he termed the 'windowpane theory' of social science, and by contesting Albert Hofstadter's (1955) separation of scientific and poetic/literary language, Joseph A. Gusfield, claimed the impossibility of any 'neutral' social scientific prose merely mirroring and reflecting the world as it is without interpretive, discursive and rhetoric distortion:

> The aim of presenting ideas and data is to enable the audience to see the external world as it is. In keeping with the normative prescriptions of scientific method, language and style must be chosen which will approximate, as closely as possible, a pane of clear glass. Scientists express their procedures, findings and generalisations in 'neutral' language. Their words do not create

or construct the very reality they seek to describe and analyse. (Gusfield 1976, 17)

According to the 'windowpane theory', scientific investigation, prose and explanation must be as dispassionate, objective, neutral and clinical as possible. Moreover, any interference with this 'windowpane' works merely as a smoke screen for unscientific or pseudo-scientific procedures and masks illegitimate claims to valid knowledge. Several scholars, however, have convincingly argued against such a positivistic 'windowpane' image of social science and defended metaphors as an integral and important part of scientific – and indeed sociological – reasoning (see Burke 1954; Rigney 2001). According to them, metaphors function as indispensable explanatory, perceptual and semantic prisms – and not as panes of clear glass – through which the complexities and intricacies of social life can be comprehended, and Alan Wolfe went as far as observing how 'it is impossible to capture the complexity and interconnectedness of human society without metaphors' (Wolfe 1993, 164).

Consequently, metaphors may perform several fundamental tasks for sociological theorising and we contend that they are particularly useful to 'think with'. Therefore, six major, and interrelated, achievements of metaphorical thinking – as part of the Burkean 'perspectives of incongruity' – in sociology can be detected: *transformation*, *transference*, *transgression*, *transportation*, *transmutation* and *transcendence*. First, metaphors are *transforming* – by their deployment, they may creatively change our conception of the actual world; they transform it from something alien to something recognizable or vice versa. Second, they are *transferring* – they apply the language or images from one domain of human experience to another, quite often in absurd but nevertheless fertile and useful fashion. Third, they are *transgressing* – they allow for a transgression and suspension of reality 'as we know it' and, momentarily, present us with a possible image of how the world could or might just as well be seen. Fourth, they are *transporting* – they take us into a wonderful world of make-believe in which things are not always as they seem. Fifth, they are *transmuting* – they enable a transmutation of ingrained academic doxa or common sense by way of refreshing perspectives or innovative and surprising juxtapositions and they, perhaps also only momentarily, reorganise and reconfigure our ideas and notions about the actual social world and its fundamental workings. Finally, metaphors may assist in *transcendence* – they expand our perceptual access *to* the real world and, perhaps, also point to ways of transcending – through practice – limitations *in* the real world. In short, armed with the magic of metaphors sociologists may hope to see further or deeper into the social texture than they would be able or allowed to without them.

Common to all of the above aspects of sociological 'thinking with' metaphors is what has been termed 'rhetorical redescription', and metaphors are indeed an integral aspect of the underlying rhetoric or poetics of sociology (Brown 1977; Edmundson 1984; Gusfield 1976; Simons 1990). Thus, Paul Ricoeur captured the centrality of metaphors for social analysts aiming at comprehending the world by stating that 'metaphor is the rhetorical process by which discourse unleashes the power that certain fictions have to redescribe reality' (Ricoeur 1977, 7). Through metaphorical processes a distinct and powerful yet fictitious 'as if' quality is imposed on social reality that – in Erving Goffman's (1959) succinct terms – functions as a temporary 'scaffold' assisting in imagining and accepting a certain structure to the otherwise complex world. Through such metaphorical processes, metaphors may either defamiliarise reality (making it appear wonderfully alien and strange) or familiarise the world, whereby it appears more recognizable and palpable. In either case, metaphors provide an arsenal of concepts, ideas, notions and hypotheses that enhances and buttresses understanding, and which points to what Lubomir Doležel (2000) aptly termed 'possible worlds'.

Moreover, metaphors as such rhetorically redescriptive devices are invoked in order to persuade readers – momentarily – that they suspend their normal quotidian perception of how things are and enter a recontextualised and reconceptualised realm of fiction. To do so convincingly, it is however necessary that certain links and connecting points are proposed and sustained which organise scattered observations and concepts into more comprehensive and compact 'metaphorical networks' (Corradi 1990). One might, with recourse to the terminology of poetry, call such interweaving metaphorical networks a 'storyline'. Thus, the power of metaphorical persuasion is to a large part embedded within their ability to make the improbable appear probable, and in them making the absurd look acceptable by way of imposing structure where apparently no structure is visible. Finally, there is an indistinguishable moral edge to most metaphors. Metaphors signal not only a concern with the rhetorical redescription of poetically probable or fictitious worlds – how the world could or might be seen – they may also depict a deep-seated moral, sometimes even political, sentiment regarding how the would should or should not be (Denham 2000). We will return to this latter issue in the conclusion.

Because they both cherish metaphors as such analytical – and to some extent also moral – tools for practicing sociologists, Zygmunt Bauman and John Urry have written extensively on and throughout their own work applied a multitude of metaphors. Urry, for example, states that 'sociological thinking, like any other form of thought, cannot be achieved non-metaphorically' (Urry 2000, 21) and Bauman passionately counters the criticism of metaphors and captures the essence of metaphorical thinking for sociological understanding:

> The desperate efforts of many a scientist to cut off all metaphorical roots and hide all traces of kinship with 'ordinary' (read: non-scientific, inferior to scientific) perception and thought are (perhaps an inevitable and certainly expectable) part of a more general tendency of science, all too-evident

since Plato commanded philosophers to venture out of the cave, to put a distance between itself and the 'common sense' of *hoi polloi* ... I deny that this means that using metaphors is a sign of a lesser and inferior knowledge. Using metaphors derives from and signals our responsibility towards the prospective human objects/participants of the activity known under the name of 'sociology' – an activity that is the sole source of whatever authority we may claim and acquire. It signals a refusal to act under false pretences, to bid for greater authority than realistically can be claimed, and above all to distort the subject-object communication (yes, communication, since both the subject and the object are human and both have tongue) in the subject's (that is, the sociologist's) favour. (Bauman in Jacobsen et al. 2007, 265)

Many different metaphors have been proposed throughout the history of sociology to account for human conduct, structural components or other social phenomena (Rigney 2001). In recent years, however, one might argue that a metaphorical shift or discursive change (or even epistemological rupture) have occurred from sedantic, stable and stationary metaphors (such as metaphors of the machine, closed system and legal order) to a focus on liquidity, flow and mobility (as is particularly evident in the literature on globalisation and in the 'new mobility paradigm'). Both Bauman and Urry have spearheaded and been instrumental in promoting this epistemological rupture, the essence of which we will now exemplify through a cursory excursion into their writings.

Zygmunt Bauman: Stratified Mobility and Polarised Globalisation

The interest in mobility came to Zygmunt Bauman primarily as part of his general development of a 'sociology of postmodernity' (note, not a 'postmodern sociology') and his increasing concern with the themes of globalisation and consumerism throughout the late 1990s. However, his preoccupation with the plight of those at the bottom of society remains a continuous presence in Bauman's work ever since the early years (Tester and Jacobsen 2005). As such, Bauman's work contains an unmistakable morally biased edge and in his description and diagnosis of the contemporary world of suffering he – categorically – sides with the sufferers (Jacobsen and Marshman 2008b). Moreover, in his diagnosis of the contemporary social landscape, Bauman's perspective is largely metaphorical. Throughout his many books, a multitude of penetrating metaphors have been proposed and refined – of types of intellectuals, of stages of modernity, of sorts of utopia, of forms of community, of varieties of nomadism, etc. – metaphors that all deal with and capture the stratified and polarised experience of modern, solid as well as liquid, life (see Jacobsen and Marshman 2008a, 2008c). Here we will dwell on a few of these metaphors pertaining to contemporary social inequality and mobility.

'Vagabond' versus 'Tourist'

Although global inequality and domination materialise in many different forms and shapes (Held and Kaya 2007), Bauman begins his analysis of the advance of 'liquid modernity' – replacing previous 'solid modernity' – by stating:

> The era of unconditional superiority of sedentarism over nomadism and the domination of the settled over the mobile is on the whole grinding fast to a halt. We are witnessing the revenge of nomadism over the principle of territoriality and settlement. In the fluid stage of modernity, the settled majority is ruled by the nomadic and exterritorial elite. (Bauman 2000, 13)

Today, we are all nomads, but our nomadism – its causes and consequences – differs radically. While some revel in the ability to move freely and without spatial limitations, others are forced to stay on the move, bound to be on the run. Contrary to our ancestors, who – according to Bauman's metaphorical arsenal – were 'missionaries' or 'pilgrims' whose life courses could be predicted, charted and indeed imitated, the nomads leave few traces behind for their successors to follow (Bauman 1993, 240–44). Although admitting that the notion of the 'nomad' is perhaps a 'flawed metaphor' for men and women cast in the liquid modern condition (Bauman 1993, 240), Bauman continues to insist that nomadism comes in many different guises thus emphasising how:

> The exact meaning of 'nomadism' varies from one social position to another and 'mobility' comes to various people in various forms ... While some people acquire an unheard of freedom to move, others are increasingly *glebae adscripti*. In a mobile world, freedom of mobility is fast becoming a most hotly coveted and contested value and a major stake in the new global as well as inner-societal stratification. (Bauman 1998b, 206)

According to Bauman, today we are – each and every one of us, but each to different degrees – either 'tourists' or 'vagabonds' (Bauman 1993, 240–44, 1998a, 77–102). Although the actual content of these apposite metaphors remains somewhat speculative and obscure, the former refers to those sections of society able to exert influence over why, when, where and how they move, whereas the latter refers to the life experience of those unable to practice mobility out of want but who are rather on the move out of need. Examining the way we 'move' in liquid modern times, Bauman observes how entry visas are increasingly being phased-out while passport control becomes ever more important. He argues that this state of affairs 'could be taken as the metaphor for the new, emergent, stratification. It lays bare the fact that it is now the 'access to global mobility' which has been raised to the topmost rank among the stratifying factors' (Bauman 1998a, 87). To Bauman, such access to mobility, like freedom (Bauman 1988), is not only merely differential – it is *stratifying* and *polarising* in a consumer society in which the 'flawed consumers' and 'human waste' are seen as unwelcome. Thus, not only the access to mobility but also the experience of mobility varies from hospitality towards the tourist to hostility towards the vagabond:

> The first travel at will, get much fun from their travel (particularly if travelling first class or using private aircraft), are cajoled or bribed to travel and welcomed with smiles and open arms when they do. The second travel surreptitiously, often illegally, sometimes paying more for the crowded steerage of a stinking unseaworthy boat than others pay for business-class gilded luxuries – and are frowned upon, and, if unlucky, arrested and promptly deported, when they arrive. (Bauman 1998a, 89)

It is Bauman's contention that although inhabiting the same global planet, tourists and vagabonds experience the predicaments of this world radically differently. To the tourists, the voluntary mobile, the world is their oyster – their guiding star is the seductive pulls of fun, pleasure and excitement. They are able to annul time and space through mobility and travel. To the vagabonds, who on the other hand are either involuntary mobile or involuntary local, the liquid modern world and its criminalisation of poverty, unpleasantly resembles a prison cell with more pushes than pulls, more keyholes than doorknobs. While the tourists have the choice to live global lives, the vagabonds are often the locals unable to travel or, perhaps even worse, destined to be stuck in confined and stuffy places.

Bauman's metaphors of 'tourist' versus 'vagabond' paint a gloomy picture of contemporary life experiences and expectancies as being worlds apart. One the one hand, they point to the sunny side of mobility as opposed to the dark side of immobility and, on the other hand, they suggest that increasingly certain spaces become 'localities of immobilisation' or 'involuntary ghettos' of confinement (Bauman 2001a) – such as refugee camps, social housing estates, prisons and dilapidated neighbourhoods – in which the poor, indolent and downtrodden are either subjected to panoptical control or left to live in states of incomprehensible human misery. These are spaces of social disintegration and degradation, of desolation, exclusion, anomie and atomisation. Bauman's point is that it is possible, indeed appropriate, to extrapolate these metaphors of 'tourist' and 'vagabond' from the exclusive realm of geographical travel and mobility to encapsulate the entire emporium of human experience.

'Escape' and 'Exit'

Alfred O. Hirschman (1970) once suggested three possible strategies for dealing with organisational and social change: 'loyalty', 'voice' and 'exit'. Whereas loyalty as well as voice seemingly belong to the bygone age of territoriality – a solid modern age of 'heavy' capitalist production and of lasting and mutual engagement between capitalists and proletarians (as mediated by the welfare state) in which one hoped to change unacceptable circumstances by raising one's 'voice' – in today's consumer society of 'light' capitalism 'exit', at least for those at the top of the social ladder, has quickly become the preferred strategy. Thus, what especially characterises the global stratum of 'tourists' is their 'exterritorial' status: they are not *of* the place in which they move, they are merely *in* it. And despite momentary fits of homesickness, the tourists keep moving on to where their desires and dreams lead them. They are, in Bauman's terms archetypal 'sensation gatherers' for whom any impediment to 'instant gratification' constitutes the worst possible nightmare. Because notions of locality and sedentarism to them reek of stagnation, monotony and immobility, between them and their fellow nomads, the vagabonds or vagrants, a gap of global proportions opens up, a chasm of mutual incomprehension stretches, whereby an experiential wedge is mercilessly driven between the 'haves' and the 'have nots'. Whereas the former travel for pleasure, the latter keep on the move in order to avoid expulsion or incarceration. And whereas the former continuously develop ever new strategies for fleeing the moral responsibility and for evading the social obligations that comes with being tied to place, the latter will have to stay put and pray – probably hoping against hope – for better times.

In a thoroughly individualised, privatised and consumerised society such as our contemporary liquid modern social world, the only sensible strategy for those able to do so is to stay on the move and avoid standing still. Therefore, Bauman asserts how 'it is exit that our political institutions ... favour' (Bauman 1995, 286) and how 'escape becomes now the name of the most popular game in town' (Bauman in Jacobsen et al. 2007, 152). He often quotes Ralph Waldo Emerson's aphorism that 'in skating over thin ice, our safety is our speed' as the epitome of tourist experience in an uncertain era. The metaphor of 'escape', and with it also of 'exit', testifies as to how:

> the people operating the levers of power on which the fate of the less volatile partners in the relationship depends can at any moment escape beyond reach – into sheer inaccessibility ... The prime technique of power is now escape, slippage, elision and avoidance. (Bauman 2000, 11)

This means that instead of engaging with and trying to solve the deep-rooted social problems of poverty, inequality and exclusion, the only interest of the tourists is in removing any possible obstacle to capitalism and with it endless consumption. Bauman's metaphors of 'exit' and 'escape' insist that the powerful in the 'post-engagement age' evade their social and moral obligation and detest any type of commitment apart from the compulsion to consume. When not taking advantage of their unlimited mobility, they retreat into private fortresses and 'gated communities' in which they avoid meeting the *mobile vulgus*, the vagabonds, roaming in the mean streets outside. Bauman frequently invokes the notion of 'absentee lordship' to describe how 'exit' and 'escape' is the new name of the game:

> The contemporary global elite is shaped after the pattern of the old-style 'absentee landlords'. It can rule without burdening itself with the chores of administration, management [and] welfare concerns ... Travelling light, rather than holding tightly to things deemed attractive for their reliability and solidity – that is, for their heavy weight, substantiability and unyielding power of resistance – is now the asset of power. (Bauman 2000, 13)

The metaphors of 'escape' and 'exit' forcefully signal a decline of the social, a polarisation of human experience and an evasion of political commitment and moral responsibility on behalf of those who are still able to make a difference in the world. So while the affluent tourists jet-off to exotic destinations, the poor – the human waste products of liquid modernity – are destined to stay grounded or travel inhumanely.

'Mismeeting'

A central component in and prerequisite for any kind of human sociality is that people meet and talk as openly, genuinely and accessibly as possible. This is especially the case in a world of mobility and a world increasingly inhabited by strangers. According to Bauman, however, the ancient art of meeting strangers is rapidly disappearing in a society in which people increasingly insulate and isolate themselves in 'gated communities' (Bauman 2006). With a phrase from Martin Buber, Bauman terms this 'mismeeting'. By invoking the metaphor of 'mismeeting', Bauman is aiming at how avoidance of social contact creates a 'realm of non-engagement, of emotional void, inhospitable to either sympathy or hostility; an uncharted territory, stripped of signposts; a wild reserve inside the life-world' (Bauman 1993, 154). As a consequence, instead of meeting each other, we engage in a series of sequestration and exclusion strategies in our everyday experience with human encounters – strategies that are not supported by hostility, but, perhaps even worse, by indifference.

Although not a necessary precondition, physical/spatial proximity and moral proximity, at least in Bauman's understanding, may mutually reinforce and support each other. Their intimate relationship, it seems, is, and perhaps irreparably, dissolved in liquid modernity. Consequently, instead of regarding others as an 'Other' – as a carrier of moral subjectivity and as a unique human being – we, at best, rather regard them as functionalities for our own life projects and, at worst, as threats, nuisances and objects to be avoided. Thus, through the deployment of the strategy Bauman labels 'effacing the face' – covering in fact a conglomeration of de-ethicalised strategies such as 'mismeetings' – any moral concern or commitment is barred from ever evolving or coagulating:

'Cut free from its anchor in another person, responsibility can now be attached to the impersonal rules of the transaction itself' (Bauman 1990, 28). Ethics, or rather morality, is now substituted by economic transactions (the 'cash nexus') as the primary mediating mechanism between people. Even in the most intimate areas of human life such as personal relationships, love and solidarity, a 'something for something' mentality advances (Bauman 2003). In a world in which no one is really irreplaceable, everybody takes on the role as peripheral extras in the lives of others.

Underpinning this mentality of mismeeting and its accompanying devaluation of human potential – or perhaps as a concrete materialisation of its increasing success – ever new spaces or rather 'non-places' (Augé 1995) designed especially for securing and supporting mismeetings proliferate in contemporary urban landscape. Such spaces are 'public but not civil' (Bauman 2001c) and the art of civility – as ancient as that of meeting – is replaced by the art of avoiding contact with others or by making encounters as perfunctory as possible. These are spaces – such as deodorised, purified and dehumanised airport lounges, shopping malls, parking lots or plazas – oozing of distance instead of proximity; of effacing the face through the techniques of civil inattention; of mismeetings instead of meetings – in short, of anti-sociality. As Bauman claims:

> The overall effect of deploying the art of mismeeting is 'desocializing' the potentially social space around, or preventing the physical space in which one moves from turning into a social one ... The techniques of mismeeting all serve to achieve this effect and to inform whoever may watch that the effect has been achieved and indeed intended. (Bauman 1993, 155)

Thus, the dissolution of social bonds, of 'the social' and eventually of 'society' as we know it, follows from the – intended and unintended – dissocialising effects of mismeeting.

As is evident from the outline above of the metaphors of 'tourist' versus 'vagabond', 'escape' and 'exit' as well as that of 'mismeeting', Bauman's diagnose is almost exclusively critical and excessively gloomy but it is also and more than anything else, aimed at locating a hatch leading to possible alterations and improvements. Thus, the unequivocal moral message contained in Bauman's metaphors of liquid modernity can be summarised as follows:

> If the idea of 'good society' is to retain meaning in the liquid-modern setting, it may only mean a society concerned with 'giving everyone a chance' and removing all impediments to taking that chance up one by one. (Bauman 2001b, 146)

Bauman's deep-seated dissatisfaction with and moral indignation at the contemporary state of affairs in liquid modern society increasingly sifting away some groups of people as either 'unworthy' or 'wasted' lives deprived of any chance of dignity or decent livelihood (Bauman 2004) is mirrored in and becomes particularly obvious through the many colourful metaphors he ultimately connects to the ability or absence of ability to live out mobility.

John Urry: 'Network Capital' and 'Meetingness'

'Network capital' and 'meetingness' are two of the most central metaphors in John Urry's mobile sociology and both are crucial for understanding the socially differentiated nature of mobilities and how mobility is crucial to the patterns of contemporary social inequalities. While the general account in *Sociology Beyond Societies: Mobilities for the Twenty-First Century* (Urry 2000) pays little attention to (some would perhaps even say neglecting) issues of inequality, power and differentiated experiences of mobility, these features are, as we will now show, central conceptual and empirical features in the recent *Mobilities* (Urry 2007) which is simultaneously an extension and a reformulation of important parts of the argument presented in the first book. The extension aspect is evident in the continuous desire to 'develop through appropriate metaphors a sociology which focuses upon movement, mobility and contingent ordering, rather than upon stasis, structure and social order' (Urry 2007, 9), while the reformulation aspect stems from a desire to develop 'a detailed analysis of just how and why mobilities make such a difference to social relations' and to 'distinguish between different kinds of mobility systems and movements' (Urry 2007, 10). Whereas the first book focused extensively on the general experience of mobility, the latter book thus incorporates a certain concern with differentiation and potentials for inequality.

'Network Capital'

As Urry suggests, 'the notion of network is also a dominant metaphor for global times, rather than say "machines"' (2003, 50). Whereas his earlier work discussed networks in largely spatial terms (and rethinking it through the metaphor of 'fluid', see Urry 2003, ch. 4) in *Mobilities, Networks, Geographies* (Larsen et al. 2006) and *Mobilities* (Urry 2007) network is conceptualised through the economic metaphor of capital thereby explicitly linking mobility to issues of social (in)equality. Here elaborating upon Pierre Bourdieu (1984) – the *capital* part – and Manuel Castells (1996) – the *network* part – Urry develops the metaphor of 'network capital' (actually consisting of two metaphors: network *and* capital).

Following the sociological tradition of portraying the 'social' through, for example, natural or biological metaphors (see Jacobsen 2007; Rigney 2001; Wolfe 1993) – e.g. seeing society as an organism as in the work of Émile Durkheim – the network metaphor, as it is developed in Castells (1996) *informational* account of the network society, represents the spatial typologies of the social as a 'spider-less spider web' structure, a threaded structure of more or less interconnected lines and hubs through which flows of information spread across distances in real time. Urry's work can be seen an *extension* of Castells's network society. Castells's account is, from the outset, a macro one, and he does not address the network society in relation to friendship and family life, for *social* networks. Yet, this notion suggests that 'sociality' is increasingly networked with mobile communication and performed through interfaces and phonescapes, rather than purely face-to-face. Much of Urry's later work has explored how social networks of family life and friends increasingly become materially and technologically networked and expand across space so that social life and one's close ties are increasingly at-a-distance.

Urry also emphasises how information and communication technologies and especially new kinds of software change the nature of business and social life. And yet, he critiques Castells for being overly cognitivist since Castells's account is one of foot-loose 'informationalism' bypassing places, face-to-face sociality and human net*working*. What is lacking in Castells's account according to Urry is the continuing significance of intermittent face-to-face meetings and increasing significance of intermittent corporeal travel to attend such meetings since network ties are increasingly at-a-distance (see next section). This means that networking and social relations depend more and more on communications and occasional travel (as demonstrated in Larsen et al. 2006). This also means *access* to communication technologies, transport, meeting places *and* the social and technical *skills* of networking is thus crucial for sustaining social networks and hence general social wellbeing. This is what Urry calls 'network capital'. In societies organised around 'circulation' and 'distance', the greater the significance of network capital within the range of capitals available within a society (Urry 2007, 52):

> Network capital is the capacity to engender and sustain social relations with those people who are not necessarily proximate and which generates emotional, financial and practical benefit ... Those social groups high in network capital enjoy significant advantages in making and remaking their social connections, the emotional, financial and practical benefits. (Urry 2007, 197)

Urry is not interested in mobility technologies in and of themselves but in what they *afford* to social networks, how they *potentially* change them:

> What are key are the social consequences of such mobilities, namely, to be able to engender and sustain social relations with those people (and to visit specific places) who are mostly not physically proximate, that is, to form and sustain networks. So network capital points to the real and potential social relations that mobilities afford. This formulation is somewhat akin

to that of Marx in *Capital* where he focuses upon the *social* relations of capitalist production and not upon the *forces* of production *per se* (1976). My analogous argument is that it is necessary to examine the social relations that the means of mobility afford and not only the changing form taken by the forces of mobility. (Urry 2007, 196)

Castells's account of the network society is, according to Urry, a one-sided perspective on 'informationalism' and communication. In contrast, Urry emphasises how social networks even in network societies are engendered and sustained through intermittent face-to-face *meetings* that increasingly require costly and time-consuming travel to take place (the key here being meetings and not travel as discussed later) and 'network capital' therefore comprises:

1. *An array of appropriate documents, visas, money, qualifications*
2. *Others (workmates, friends and family members) at-a-distance*
3. *Movement (and communication) competences*
4. *Location free information and contact points*
5. *Communication devices*
6. *Appropriate, safe and secure meeting places*
7. *Physical access* to cars, roadspace, fuel, lifts, aircraft, trains, ships, taxis, buses, trams, minibuses, email accounts, internet, telephones and so on
8. *Time and other resources to manage and coordinate no. 1–7.* (Urry 2007, 197–8)

Thus, 'network capital' complexly comprises technical, cognitive and social skills and depends on 'accesses' to various documents, strong and weak ties, communication technologies, databases *and* transport technologies. While some of these 'accesses' depend upon appropriate 'economic capital', others are not necessarily economic in nature. For instance, many 'poorer' migrants are likely to have many 'others' at a-distance (2). Likewise, while 'access' to communications and travel requires 'economic capital', what Urry terms 'competences' (3) is largely independent of this type of capital. So there is no linear or direct proportionality between 'economic capital' and 'network capital'. And as people are distributed 'far and wide', so 'network capital' is essential for social life: 'Without sufficient network capital people will suffer social exclusion since many social networks are more far-flung' (Urry 2007, 179).[1] However, while there seems to be a general trend towards increasingly distanciated social networks not everyone has network ties predominately at

---

1 The metaphor of 'network capital' resembles Kaufmann et al'.s notion of 'motility' that refers to *potentials* for actual mobility. Motility 'encompasses interdependent elements relating to access to different forms and degrees of mobility, competence to recognise and make use of access, and appropriation of a particular choice, including the option of non-action' (Kaufmann et al. 2004, 750).

a-distance and this capital is obviously of more value to some groups than others (something *not* discussed by Urry). Migrants and Diaspora groups appear to be highly depended upon 'network capital' and yet they are often *forced* to network at-a-distance, sometimes with very little 'network capital'. And partly therefore they have developed striking 'competences' in using communications (see Larsen and Urry 2008 as well the anthropological literature on ICT and migrants, e.g., Horst 2006; Parreñas 2005; Vertovec 2004; Wilding 2006). Urry accepts Bauman's (1998a) claim that mobility and speed has climbed to the top of the stratification hierarchy, the reason for this being the unequal distribution of 'network capital' among various groups and nations. In other words, Urry suggests that the metaphor of 'network capital' is the complex yet systematic tool for understanding how the unequal distribution of *potentials* for networking at-distance have become the major stratifying factor in contemporary societies.

Urry compares *and* contrasts 'network capital' with Robert Putnam's notion of '*social* capital' that 'refers to connections among individuals – social networks and norms of reciprocity and trustworthiness that arise from them' (Putnam 2000, 19; Urry 2007, 198–9). There are striking similarities between these two capitals because sustaining *connections* is the key in both concepts. The difference relates to whether cultures of mobility destroy or ensure such social connections. To put it short, Putnam perceives connections as being fostered within propinquitous communities and cultures of mobility are seen as a threat to such propinquitous communities (Putnam 2000, 407–8). In contrast, Urry argues that in increasingly mobile and networked societies mobilities are often necessary for the production of 'social capital'. 'Social capital' stretches over long geographical distances if there is appropriate 'network capital'. When network ties are not present in propinquitous communities, regular letters, packets, photographs, emails, text messages, money transactions, telephone calls and intermittent meetings are means through which love and caring take place. The concept and metaphor of 'network capital' brings out the way in which co-presence and trust can be generated at-a-distance and it presupposes extensive and predictable travel and communications. So 'network capital', according to Urry, potentially enables disconnected people to produce 'social capital', that is, to connect and is therefore a socially engendering phenomenon.

'Meetingness'

We have seen how Bauman uses the metaphors of 'mismeeting', 'escape' and 'exit' to highlight how mobile power is unequally exercised in liquid modernity. While Urry also stresses in passing how 'network capital' can be used to exercise 'exit', to avoid social obligations and unwanted sociality, his argument is rather that those high in 'network capital' are privileged and powerful because this capital is exchangeable for what he metaphorically calls 'meetingess'. 'Meetingness' is valuables because any network according to Urry:

> ... only functions if it is intermittently 'activated' through occasioned co-presence from time to time. *Ceteris paribus*, 'network activation' occurs if there are periodic events each week, or month or year, when meetingness is more or less obligatory ... In order to continue to be within a given network there are obligations to travel, to meet up and to converse. (2007, 346)

We witness an increasing amount of corporeal travel because travel enables distanciated significant others to have pleasurable, yet obligatory face-to-face meetings.

So in contrast to 'the-end-of-the-social' portrait painted by Bauman, Urry argues that we travel more and more not to escape and disconnect but to connect with significant others and fulfil obligations that can not be satisfactorily fulfilled through communications. While sporadically speaking of business meetings, Urry mainly refers to 'meetingness' in relation to friends and families more or less informally visiting each other or meeting up in significant places, so a *business* metaphor is used to explain the nature of affective sociability such as family meals, birthdays, weddings, friendship reunions and so on (see Larsen et al. 2006; Urry 2007). His point is not that 'visiting' today resembles the instrumentality and formality of businesses meetings but rather that friends and family members increasingly only meet up *intermittently*, that such visiting requires much coordination affected through communications and that travel is required for attending since many network ties are not just around the corner but at-a-distance. So in contrast to Castells, Urry argues that old fashioned proximate, multi-sensuous face-to-face sociality is still crucial in network societies. Urry's positions owes much to Georg Simmel, Erving Goffman and especially Deidre Boden and Harvey Molotch who once maintained that in writing on 'time-space compression', 'the robust nature and enduring necessity of traditional human communication procedures have been underappreciated' (Boden and Molotch 1994, 258). Boden and Molotch speak of a 'compulsion to proximity' and Urry argues that 'much travel demand seems to stem from a powerful 'compulsion to proximity', to feel the need to be physically co-present and to fulfil social and cultural obligations with significant others (sometimes against one's will ...)' (Larsen et al. 2006, 5). Face-to-face meetings, with their rich world of eye contact, facial gestures, body language, 'face-work', trust-building, intimacy, affective and pleasurable sociability and so on, is one crucial aspect of 'network capital' and given that such meetings require co-presence and travel, they are both time-consuming and costly compared to most mediated networking. Intermittent face-to-face meetings sustain social life that involves much virtual travel and communicative travel in the long periods of distance and of solitude. In mobile networked societies with connections at-a-distance and people being less likely to bump into or visit their strong ties on a daily or weekly basis, the potential for undertaking physical travel for intermittent meetings is a key for meeting the social obligations that are involved in being part of a social network. So far phones and various types of screens have been poor substitutes for the sensuous richness of face-to-face sociality. Transport and meetings at-a-distance are increasingly *necessary* and *obligatory* to networked social life, and yet many individuals are more or less excluded from participating in such network meetings because they are time-consuming and not least expensive.

Urry's metaphors of 'network capital' and 'meetingness' highlights how social life is increasingly networked and at-a-distance, and how access to and skills of performing communications, transport and intermittent 'meetingness' has become 'necessary' not least for the production of 'social capital'. The challenge Urry puts forward to transport researchers, as well as others concerned with the changing social landscape, is to understand the complexity of 'network capital' and understand the interdependencies between various forms of corporeal, virtual and communicative travel. The policy implications of all this is:

> Thus, if all else were equal, a 'good society' would not limit travel, co-presence and resulting good conversations. Such a society would extend the capabilities of co-presence to every social group and regard infringements of this as undesirable. (Urry 2007, 311)

And yet the problem is that such a good society of equally distributed mobility is likely to be a 'bad society' from an environmental, global heating, perspective. In other words, what might be socially sustainable is likely to be environmentally unsustainable. Urry's radical solution to this dilemma is to suggest that policies ought to secure that the mobile elite (and this will include many academics!) travel much less so that the 'mobility poor' can travel more:

> [T]his 'contact' should be available for all social groups at least from time to time and not just to those who are currently easily able to meet 'face-to-face'. A socially inclusive society would elaborate and extend the capabilities of co-presence to all its members. It would minimise 'coerced immobility'. Initiatives in transport, planning and communications should promote networking and meetingness (and limit missingness). (Urry 2007, 312)

As is evident here, Urry has gradually begun to add an unmistakable moral dimension to his metaphors of mobility and meetingness by invoking notions of 'the good society' and a 'socially inclusive society'. In this way, a certain convergence between the metaphors of Urry and Bauman may be detected.

Metaphors for Mobile Modernity

John Urry rightly suggested that 'the development of a 'mobile sociology' demands metaphors that do view social and material life as being "like the waves of a river"' (Urry 2003, 59). As Bauman and Urry shows us, and as Daniel Rigney rightly contended in his *The Metaphorical Society*, 'it is an impaired soul indeed that cannot appreciate the music of metaphor and the architecture of analogy' (Rigney 2001, 209). Urry and Bauman are both exponents of such a mobile and metaphorical sociology, but, as we have shown, while their metaphors of mobile life and mobile inequalities to some extent overlap, they also vary considerably. One reason for this variation is that Bauman's perspective is explicitly critical, morally inspired and humanistic; it is concerned with the *human* consequences of global mobilities and especially those that are affected negatively by them, those that belong to the dark side of immobility or forced to 'vagabondish' mobility. Indeed, the metaphorical pairing of 'tourist' and 'vagabond' is rhetorically transformative and transgressive because it simultaneously illuminates the sunny side and dark side of 'mobile modernity'. Overall, Bauman's metaphors spell dissolution, disengagement and distance and they signal the rise of new inequalities and stratifying mechanisms on a global scale. In short, social exclusion is matched and indeed promoted by spatial and physical exclusion. Bauman's general idea of 'negative globalisation' points to the fact that globalisation – and with it the pushes and pulls of mobility – is far from an unequivocal joyride for all involved parties.

In comparison to Bauman, Urry's perspective is more descriptive, aloof, systematic and almost 'post-humanistic'. While not blind to inequalities, Urry does not take sides with the 'mobility poor' and much of his theorising and research is concerned with the sunny side of mobile modernity, of airports, tourism, business meetings and so on. While we have seen that Urry has gradually begun to include discussions of 'the good society', his sociology is nonetheless largely amoral, focusing upon what 'is', and with making a detached grand sociology of this new 'mobile modernity'. And whereas Bauman is concerned with humans of flesh and blood, enjoying or suffering, following the tenets of complexity theory and especially Actor-Network-Theory (that is inherently critical of critical theory), Urry's perspective is a 'post-humanistic' one in which systems, objects and machines are as important as human beings, and 'humans' only exist in conjunction with such inhuman components. We may say that whereas Bauman is concerned with the (lack of) *human-ness* of mobile modernity, Urry's lens is that of describing, in part through appropriate metaphors, its sys*tem-ness* (a much used metaphor in his book *Mobilities*).

The metaphors of 'mismeeting' and 'meetingness' illustrate their differences, in part because they are in fact *not* mutually exclusive, competing metaphors. With the metaphor 'mismeetings', Bauman stresses that our age is one of mobile 'gated communities' where strangers seldom meet and the 'elite' only meet likeminded souls in designated meeting spaces at-a-distance. But the latter, meetings between like-minded at-a-distance, is precisely what Urry describes with his metaphor of 'meetingness'. Whereas Bauman gloomily stresses the end-of-the-social, Urry is keener to highlight how social networks are undergoing transformations in mobile network societies, with new sets of social obligations and forms of caring, of communicating, travelling and meeting up, something largely neglected by Bauman (however Urry is rightly gloomy when it comes to the unintended environmental consequences of mobile modernity). Another problem with Bauman's 'tourist' metaphor is that it wrongly reduces the significance of tourist travel to questions of desires and pleasures, and in this depiction tourism becomes synonymous with something superficial and unnecessary. In contrast, Urry has increasingly argued and shown that in mobile societies where networks are distanciated, much tourist travel is concerned with obligatory 'meetings' (e.g. birthdays, weddings, reunions) within families and friendship groups living elsewhere, making tourist travel a 'necessary' component of 'network capital' and by implication of 'social capital' (Larsen et al. 2006).

A related problem with Bauman's metaphors is that they tend to polarise the world with their dual worlds of 'mobility rich' and 'mobility poor', 'tourists' and 'vagabonds', thereby resembling some sort of binary thinking. George Steiner (1971) once insisted that today dialectics have become binary whereby dialectical transcendence, one of the six aforementioned characteristics of metaphors, is lost. It seems that Bauman's metaphors commit such a binary fallacy as his metaphors may solidify the world and our understanding of it instead of allowing for transmutation and transcendence. A related problem is that metaphors stick – and especially Bauman's metaphors have inspired the imagination of many students of scholars of mobility. Metaphors, when applied uncritically or disseminated widely, however tend to become reified and the danger lurks of so-called 'sociological semanticide' through reification, tautology and the killing of semantic innovation (Turner and Edgley 1980). It seems to us that the metaphor of the 'migrant' could quite possibly complement and transcend the inherently binary logic of 'tourist' and 'vagabond'. Migration is far from a one-way journey of leaving one's homeland behind, but often a two-way 'journey' or circulation between two sets of 'homes' and with 'national' network members in various other countries. The metaphor of the 'migrant' helps highlighting how many people, from very different walks of life, 'live', feel at home and network in more than one place across nations (in contrast, Bauman's 'tourist' always travels to new places and return to the same home while the vagabond is 'homeless').

Although we acknowledge the creative importance of metaphors which do not arise from actual empirical research, but are the product of academic contemplation, following Urry we suggest that metaphors must ultimately prove their scientific productivity, usability and plausibility in relation to other metaphors, theories, findings and in actual empirical research (Urry 2000, 27). This may lead to a modification of metaphors, or even their rejection, and the subsequent development of new ones. Thus, we do not argue for rigid positivistic testing of such metaphors but rather for their creative use – as imaginative tools – in actual research of inequalities on the move, of inequalities that are not contained within the nation state, of mobility that is constantly changing. If we are to employ such metaphors empirically, it is crucial to dispense with, or at least extend sociology's traditional perspective of 'methodological nationalism' with a more 'cosmopolitan outlook' (Beck 2007) or mobile methods (Urry 2007) that recognises that power and social (in)equalities increasingly relate to possibilities for communicating and moving across national borders in safe, pleasant, fast and predictable fashions as more and more people, of both those high and low in 'network capital', life transnational lives or are affected hereby. In particular, we argue for the need of researching ethnographically – but aided and abetted by metaphors – how such mobile groups network in practice, with little or much 'network capital'. Studies of social (in)equality needs to move beyond registering differences in actual and potential 'network capital' and also explore ethnographically – as well as metaphorically – how network capital is put into practice in concrete everyday life.

References

Augé, M. (1995), *Non-Places: Introduction to an Anthropology of Supermodernity* (London: Verso).
Bauman, Z. (1988), *Freedom* (Buckingham: Open University Press).
Bauman, Z. (1990), 'Effacing the Face: On the Social Management of Moral Proximity', *Theory, Culture and Society* 7:1, 5–38.
Bauman, Z. (1993), *Postmodern Ethics* (Oxford: Blackwell).
Bauman, Z. (1995), *Life in Fragments* (Oxford: Blackwell).
Bauman, Z. (1998a), *Globalization: The Human Consequences* (Cambridge: Polity Press).
Bauman, Z. (1998b), 'Identity – Then, Now, What For?', *Polish Sociological Review* 123:3, 205–16.
Bauman, Z. (2000), *Liquid Modernity* (Cambridge: Polity Press).
Bauman, Z. (2001a), *Community: Seeking Safety in an Insecure World* (Cambridge: Polity Press).
Bauman, Z. (2001b), 'The Great War of Recognition', *Theory, Culture and Society* 18:2–3, 137–50.
Bauman, Z. (2001c), 'Uses and Disuses of Urban Space', in Czarniawska, B. and Sollim, R. (eds).
Bauman, Z. (2003), *Liquid Love* (Cambridge: Polity Press).

Bauman, Z. (2004), *Wasted Lives: Modernity and Its Outcasts* (Cambridge: Polity Press).
Bauman, Z. (2006), *Liquid Fear* (Cambridge: Polity Press).
Beck, U. (2007), 'Beyond Class and Nation: Reframing Social inequalities in a Globalizing World', *British Journal of Sociology* 58:4, 679–705.
Boden, D. and Molotch, H. (1994), 'The Compulsion of Proximity', in Friedland, R. and Boden, D. (eds), *Nowhere: An Introduction to Space, Time and Modernity* (Berkeley, CA: University of California Press), 257–86.
Bourdieu, P. (1984), *Distinction: A Social Critique of the Judgment of Taste* (London: Routledge and Kegan Paul).
Brown, R.H. (1977), *A Poetic for Sociology* (Cambridge: Cambridge University Press).
Burke, K. (1954), *Permanence and Change: An Anatomy of Purpose* (Los Altos, CA: Hermes Publications).
Castells, M. (1996), *The Rise of the Network Society* (London: Blackwell).
Castells, M. (2001), *The Internet Galaxy: Reflections on the Internet, Business and Society* (Oxford: Oxford University Press).
Castells, M., Fernandez-Ardevol, M., Qiu, J.L. and Sey, A. (2007), *Mobile Communication and Society: A Global Perspective* (Cambridge, MA: MIT Press).
Corradi, C. (1990), 'The Metaphoric Structure of Scientific Explanation', *Philosophy and Social Criticism* 16:3, 161–78.
Cresswell, T. (2006), *On the Move* (London: Routledge).
Denham, A E. (2000): *Metaphor and Moral Experience* (Oxford: Clarendon Press).
Doležel, L. (2000), *Heterocosmica: Fiction and Possible Worlds* (Baltimore, MD: Johns Hopkins University Press).
Edmundson, R. (1984), *The Rhetoric of Sociology* (London: Macmillan).
Goffman, E. (1959), *The Presentation of Self in Everyday Life* (Harmondsworth: Penguin Books).
Gusfield, J.A. (1976), 'The Literary Rhetoric of Science: Comedy and Pathos in Drinking Driver Research', *American Sociological Review* 41:1, 16–34.
Harvey, D. (1989), *The Condition of Postmodernity* (Oxford: Blackwell).
Held, D. and Kaya, A. (eds) (2007), *Global Inequality* (Cambridge: Polity Press).
Hesse, M.B. (1966): *Models and Analogies in Science* (Notre Dame, IN: University of Notre Dame Press).
Hirschman, A.O. (1970), *Exit, Voice, and Loyalty: Responses to Decline in Firms, Organizations and States* (Cambridge, MA: Harvard University Press).
Hofstadter, A. (1955), 'The Scientific and Literary Uses of Language', in Bryson, L. et al. (eds).

Horst, A.H. (2006), 'The Blessings and Burden of Communication: Cell Phones in a Jamaican Transnational Social Field', *Global Networks* 6:2, 143–59.

Jacobsen, M.H. (2007), 'Sociologiens metaforiske samfund – metaforer om 'samfundet', 'samfundet' som metafor', *Sosiologisk Tidsskrift* 15:4, 285–316.

Jacobsen, M.H. and Marshman, S. (2008a), 'Bauman on Metaphors – A Harbinger of Humanistic Hybrid Sociology', in Jacobsen, M.H. and Poder, P. (eds), *The Sociology of Zygmunt Bauman: Challenge and Critique* (Aldershot: Ashgate).

Jacobsen, M.H. and Marshman, S. (2008b), 'The Four Faces of Human Suffering in the Sociology of Zygmunt Bauman – Continuity and Change', *Polish Sociological Review* 161:1, 3–24.

Jacobsen, M.H. and Marshman, S. (2008c), 'Bauman's Metaphors – A Case Study in the Creative Use of the Sociological Imagination', *Current Sociology* 56:5, 798–818.

Jacobsen, M.H., Marshman, S. and Tester, K. (2007), *Bauman Beyond Postmodernity* (Aalborg: Aalborg University Press).

Jokinen, E. and Veijola, S. (1997), 'The Disorientated Tourist: The Figuration of the Tourist in Contemporary Cultural Critique', in Rojek, C. and Urry, J. (eds), *Touring Cultures: Transformations of Travel and Theory* (London: Routledge).

Kaufmann, V., Bergman, M.M. and Joye, D. (2004), 'Motility: Mobility as Social Capital', *International Journal of Urban and Regional Research* 28, 745–56.

Kenyon, S., Lyons, G. and Rafferty, J. (2006a), 'Transport and Social Exclusion: Investigating the Possibility of Promoting Inclusion through Virtual Mobility', *Journal of Transport Geography* 10:3, 207–19.

Lakoff, G. and Johnson, M. (1980), *Metaphors We Live By* (Chicago, IL: University of Chicago Press).

Lakoff, G. and Turner, M. (1989), *More Than Cool Reason – A Field Guide to Poetic Metaphor* (Chicago, IL: University of Chicago Press).

Larsen, J., Urry, J. and Axhausen, K.W. (2006), *Mobilities, Networks, Geographies* (Aldershot: Ashgate).

Larsen, J. and Urry, J. (2008), 'Networking in Mobile Societies', in Baerenholdt, J.O. and Granas, B. (eds), *Mobility and Place. Enacting North European Peripheries* (Aldershot: Ashgate).

Lethbridge, N. (2002), *Attitudes to Travel* (London: ONS).

Massey, D. (1991), *Space, Place and Gender* (Cambridge: Polity Press).

Massey, D. (2002 [1976]), *Sociology as an Art Form* (New Brunswick, NJ: Transaction Publishers).

Papastergiadis, N. (1999), *The Turbulence of Migration: Globalization, Deterritorialization and Hybridity* (Cambridge: Polity Press).

Parreñas, R. (2005), 'Long Distance Intimacy: Class, Gender and Intergenerational Relations between Mothers and Children in Filipino Transnational Families', *Global Networks* 5:4, 317–36.

Putnam, R. (2000), *Bowling Alone* (New York: Simon and Schuster)

Ricoeur, P. (1977), *The Rule of Metaphor* (London: Routledge and Kegan Paul).

Rigney, D. (2001), *The Metaphorical Society: An Invitation to Social Theory* (Lanham, MD: Rowman and Littlefield).

Scholte, J.A. (2005), *Globalization: A Critical Introduction* (London: Palgrave/Macmillan).

Simons, H.W. (ed.) (1990), *The Rhetorical Turn: Invention and Persuasion in the Conduct of Inquiry* (Chicago, IL: University of Chicago Press).

Steiner, G. (1971), *In Bluebeard's Castle: Some Notes Towards the Re-Definition of Culture* (London: Faber).

Tester, K. and Jacobsen, M.H. (2005), *Bauman Before Postmodernity: Invitation, Conversations and Annotated Bibliography 1953–1989* (Aalborg: Aalborg University Press).

Turner, R.E. and Edgley, C. (1980), 'Sociological Semanticide: On Reification, Tautology and the Destruction of Language', *Sociological Quarterly* 21, 595–605.

Urry, J. (2000), *Sociology Beyond Society: Mobilities for the 21st Century* (London: Routledge).

Urry, J. (2003), *Global Complexities* (Cambridge: Polity Press).

Urry, J. (2007), *Mobilities* (Cambridge: Polity Press).

Vertovec, S. (2004), 'Cheap Calls: The Social Glue of Migrant Transnationalism', *Global Networks* 4:2, 219–224.

Wilding, R. (2006), 'Virtual 'Intimacies'? Families Communicating Across Transnational Contexts', *Global Networks* 6:2, 125–142.

Wolfe, A. (1993), *The Human Difference* (Berkeley, CA: University of California Press).

PART II
Empirical Applications

This page has been left blank intentionally

Chapter 5
Mobilities and Social Network Geography: Size and Spatial Dispersion – the Zurich Case Study

Andreas Frei, Kay W. Axhausen and Timo Ohnmacht

Introduction

During the last few decades, we have witnessed enormous changes in the movement of goods, information and people. Communication techniques and transport potential have been considerably expanded and general costs of travel and communication have been substantially reduced. Due to these developments, diversity of choice has been widely expanded and exploited, producing interaction on a national and global scale through almost universal communication and transport networks. Thus, it can be argued that the world is shrinking – physically, socially and imaginatively – due to a substantial increase in accessibility through various transportation and communication systems (Axhausen et al. 2008). On a micro-societal level, social, political, economic and cultural opportunities – and enforcements – encourage us to be 'on the move' (Urry 2008a). Ongoing market globalisation with increased international competition, changing demographic structures, values, attitudes and expectations has allowed new spatial settings to evolve.

All these factors have contributed to the emergence of a new research field that is now discussed in joint projects between transport researchers, sociologists and geographers. This new research field is based on basic principles of mobility in modern life, and the related interplay between the size and structure of social networks with new forms of communication and transport (Ohnmacht et al. 2008; Larsen et al. 2006; Carrasco and Miller 2006).

In this chapter, we first give a brief overview of relevant issues in both social science and transport planning as they relate to size and structure of social network geographies. We evaluate debates about patterns of inequality structures (for a conceptual discussion of mobilities and inequality see Ohnmacht, Maksim and Bergman in this volume). Second, we focus on methodological challenges to survey data on social networks and personal mobilities. Third, we examine the size and spatial spread of social contacts in Zurich, Switzerland.[1] Based on what we believe to be the largest new quantitative survey on egocentric social networks and personal mobilities, we will try to answer the following questions:[2] what size and spatial dispersion do the social networks have and how are our findings related to patterns of inequality structures? Fourth, we conclude by summarising our main empirical findings against the patterns of inequality and suggest further research questions.

Mobilities and Social Networks

This section consists of a brief overview on relevant mobilities and social networks literature. We present theoretical debates in both transport studies and social sciences giving an overview of this relatively new research field. We will focus primarily on spatial distribution, the main topic in the empirical part of this contribution. We then consider the relevance of this research field against the background of contemporary debates on social inequality.

Mobilities and Social Networks in Social Science

Recent research puts relations between society and space at the centre stage of social theory. Since Beck (2008); Urry (2008a); Sheller and Urry (2006); Kaufmann et al. (2004) proclaimed a new 'mobilities paradigm' in social sciences, the analysis of spatial distance and social processes is no longer just an issue of transport studies and transport geography:

> Most social science has not seen distance as a problem or even as particularly interesting (except for transport studies and transport geography). This mobilities paradigm, though, treats distance as hugely significant, as almost the key issue with which social life involving a complex mix of presence and absence has to treat). (Urry 2008b, 19)

Due to this 'spatial turn' in social science, geographical space is no longer seen only as a passive container. Interestingly, geographical space was already considered relevant in early sociological theory, e.g. in work by Simmel (1908), who ascertained that spatial distance determines social proximity. But before we can re-think physical space in relation to the ordering of social relations, it

---

1 The work here was possible because of the generous and gratefully acknowledged support of the Institut für Mobilitätsforschung (IFMO), Berlin and the Swiss Bundesamt für Bildung und Wirtschaft as part of the COST Action 355.

2 The closest comparable study was undertaken independently at about the same time in Toronto by a team involving sociologists and transport planners (Carrasco et al. 2008).

is necessary to understand how space has been defined in current social science debates. Inspired by German studies, Löw (2001) has developed geographical space as a key concept in sociology. She points out that space should to be understood as a socially constituted relational concept, shaped by its social, political, cultural and economic conditions. On a micro-sociological level, people are actively integrated, constructing space using various ordering processes, thus defining space as a 'relational ordering of living beings and social goods' (Löw 2005, 2). This definition assumes that every social formation produces or reconstructs its specific 'social spatialisation' (Shields 1991, 31) which is largely built up by both the living places and locations of social contacts. Social contacts, with their spatial arrangement, form and reform the social network geography. This phenomenon can also be interpreted to some extent as activity space, since intermittent visits are necessary to maintain social contacts in most cases (Urry 2008a).

Contemporary trends, such as globalisation, transnationalisation, worldwide markets with increased international competition and changing demographic structures with different values, attitudes and expectations produce new social relations spatial patterns. Therefore, we hypothesise that spatial locations' arrangements of social network members are likely to become more remote and dispersed. Numerous studies explore the increasing possibilities for organising one's life around workplace location, place of education, partner's residence, etc. Thus, it is logical to argue that 'travel distances between members of social, familial and work-related networks have substantially increased since the 1950s; on average, social networks are more spread out and less coherent' (Cass et al. 2005, 545). Spatially distant social relations mean that people have to travel long distances to meet, and need to plan their social activities further ahead of time, possibly weakening their social network, an important source of social capital.

The impact of personal mobilities on social networks needs to be measured. Numerous researchers have conducted empirical research on physical space and social impact, such as the challenge of maintaining contacts at a distance (Latane et al. 1995; Butts 2003), social proximity in immediate surroundings (Blake et al. 1956; Caplow and Foreman 1950; Snow et al. 1981), modes of transportation and communication usage and arrangement of social network geography (Larsen et al. 2006) and life cycle and arrangement of social network geography (Ohnmacht et al. 2008; Sommer 1996).

In recent debates, the concept of 'mobility biography' has become very important in discussing the development of both geographical social space and travel behaviour. In Prillwitz et al. (2006), the term mobility biography refers to a set of an individual's longitudinal trajectories in the mobility domain. It assumes the existence of events at certain moments in an individual's life that change their travel patterns to an important degree, e.g. relocation, car ownership, or other mobility characteristics (also Beige and Axhausen 2006; Scheiner 2007). Empirically, changes that have occurred over the life cycle can be portrayed as events in a mobility biography. According to Hine and Grieco (2003, 301), '[l]ife cycle stages have a consequence for mobility and accessibility'. Consequently, life cycle stages (in this understanding certain events in a mobility biography) have an effect on 'social accessibility' (Handy and Niemeier, 1997; Götz, 2007).

In summary, we hypothesise that different degrees of personal mobilities may lead to a certain 'spatialisation' of social relations, explainable to a certain degree by mobility options exercised in a person's life trajectory, here understood as a mobility biography. Further examination is necessary on spatial dispersion of relationships for social groups involved in numerous geographical changes due to education, job, etc. (Ohnmacht et al. 2008), and whether the different personal networks overlap less (Wellman 1996). A further interesting research topic concerns spatially dispersed social networks' risk that relationships become weaker, more transient and therefore more coupled to a contemporary life cycle due to increased distance (Latane et al. 1995; Butts 2003).

Transport Planning and Social Network Geography

Transport planning aims to understand, describe and model the travel choices that people make during their daily lives, including frequent journeys outside their daily activity space (Schönfelder and Axhausen 2003). Over the last few decades, the motives and determinants of individual travel behaviour have been analysed from different perspectives. While the main approach explains personal mobility due to the travellers' socio-demographics and the generalised costs of travel, travel behaviour research has added several sociologically driven analysis directions such as role patterns, household interactions, time budgets, activity planning, life and mobility styles, etc. Another factor in understanding travel behaviour is social activity travel due to physical absence of significant others. This is particularly true for leisure travel, which dominates the travel market in terms of miles travelled and trips undertaken and is primarily motivated by the need to be with others or meet them in person (Axhausen et al. 2007; Larsen et al. 2006).

Recognising the ongoing pluralisation and differentiation within western societies and the increasing degrees of social and geographical mobility, it is crucial to investigate social realities to understand travel behaviour in greater detail. As part of these developments, it is essential to understand the geography of travellers' social networks if one wants to understand destination choices. If travel is generally about meeting others, then it is important to ascertain the starting point of trips, the meeting point, to know whether travellers are informed about the opportunities offered by a destination, and to know which constraints limit their choices or availability and therefore the choices of the full group meeting. Unfortunately, researchers in sociology have had no reason to characterise and measure social networks' geographies until now (cf. Larsen et al. 2006). They have focused mainly on the structure of social networks and their impact on the social processes (Wasserman and Faust 1994). Information obtained about the locations of members in complete or egocentric networks is spatially rough, if available at all. Geographers have generally ignored this issue, so transport planners have recently undertaken new surveys to satisfy their information needs, while drawing on the extensive sociological experience in the capturing of egocentric networks (see, for example, Marsden 2005).

Social Network Geography and Inequality

In the following section, we examine how the social dimension of inequality is rooted in social network geography, especially in the interplay between space and mobilities. New mobilities regimes have serious consequences for people's lives. In fact, differential access to mobility tools generates new kinds of social inequality. Much movement is to effect face-to-face contacts with significant others. Because mobility is often a matter of obligation and burden, one must ask how and whether people can maintain their social networks over distance. Recently, social networks have been studied to determine the relevance of mobilities in maintaining social contacts (Larsen et al. 2006).

With special emphasis on intimacy and help, Schubert (1994) focuses on spatial distance between significant others. Several forms of private care for the elderly within family networks are linked to a need for proximity, because sometimes family members are forced to be close to elderly members needing care. Within family research, the nuclear family has traditionally been considered as a spatial dense network. However, in recent years, the notion of 'modified extended families' has arisen, encompassing nuclear families living at a distance. Given that options have increased substantially in postmodernity, 'Gemeinschaft' (Tönnies 1991) is not necessarily local. Thus, distance has altered helping structures. Findings in Germany indicate that the older people get, the more they rely on help from their relatives (Schubert 1994, 230). In addition, people who need help are more likely to maintain kinship contacts in their immediate spatial surrounding (neighbourhood, community etc.), particularly when they live in rural areas (Schubert 1994, 232).

In order to explore the interaction between physical distance and intimacy, Holmes (2004) examines couples living in distance relationships. Whereas the most common reasons for long-distance distant relationships in the past were war and seafaring, today's pattern is driven by the demand for spatial flexibility required for careers (e.g. double-career couples). Holmes (2004) argues that today these 'Living Apart Together' (LATs) lives are not only dictated by men's work, but also women's desire and possibilities to follow a career, attributable to changes in gender roles. To maintain relationships, one has to interact face-to-face, using mobility tools, such as public transport, season tickets, driver's licence and so on. To tackle the problem of long-distance relationships, one has to rely on large amounts of time and monetary resources, both of which relate highly to patterns of social inequality. Thus, mobilities contribute highly to social inclusion and travel is a key activity in obtaining physical proximity to significant others.

Grieco et al. (2001) reveal that the socially excluded are not clustered together spatially; instead they are scattered, sometimes over a large area. This is particularly the case for diasporic cultures, such as migration groups. Mobilities are necessary to fulfil social obligations, such as weddings, funerals, stag nights and so forth (Beck-Gernsheim 2007). For instance, Hine and Grieco (2003) focus explicitly on the 'scattered' and 'clustered' arrangements of partner, spouse, friends, relatives and so forth. They note that, especially for spatially dispersed social networks, the use of information and communication technologies (ICTs) is very important in maintaining social contacts at a distance. As for actual transport, recent literature on the issue of transport-related social exclusion provides little information on the importance of adequate transport to visit family, friends and other relevant persons. Walking and public transport are very important for meetings, while place-to-place tangential connections to spatially distant others often require cars, long-distance trains and planes. For maintaining spatially distant contacts, economic and time constraints are key factors in determining the scheduling of time-space interaction. Thus, difficulties develop for low-income and time restricted groups in attempts to be proximate with friends, family, relatives, etc. Hine and Grieco (2003, 303) argue that 'transport researchers and transport policy makers have been insufficiently focused on the consequences of a networked society for the total reorganisation of transport and travel'. On a policy level, the 'social sprawl' problem can be solved by providing all-access public transport cards to people at risk of social exclusion. This may enable social inclusion facilitated by transport as a counterbalance to dispersed social networks; e.g. high access to mobility through low generalised costs can increase opportunities to participate in society through social inclusion journeys.

We have discussed different types of transport and sociological research concerned with social networks and mobilities. From these discussions, we draw the following conclusions: changes in social network space through modernisation have been identified as interplay between significantly expanded transport and communication systems, lowered generalised costs and changing social practices. These dynamics lead to new spatial network patterns. The patterns must be addressed in greater detail for a deeper understanding of social-activity travel and social processes in general. Overall, we concluded that it is still necessary to study social relations and space together in order to understand travel in a more social way. We must focus on the question of how 'social networks involve diverse connections, which are more or less at a distance, more or less intense, and more or less mobile' (Larsen et al. 2006, 3) against the background of physical travel to fulfil social contacts. Focusing on the absence of social contacts is a chance for mobility and travel research to forge new insights into the dynamics of modern life and their effects on mobilities. For the remaining part of the chapter, we are mainly interested in two issues; first, discussing methodological challenges to research mobilities and social networks and second, examining the structure of social network geographies in Zurich, Switzerland.

Methodological Challenges: Mobilities and Social Networks

When focusing on mobilities and social networks and their mutual dynamics, it is necessary to develop and apply 'mobile field methods' in data gathering to explore and examine recent dynamics in greater detail. This section explains the Zurich survey, how it was conducted and how representative the sample obtained was. We follow with a discussion about the name generator used in the survey to help respondents name their social contacts. In addition, we highlight the methodological challenge of measuring spatial spread of a social network.

Survey

Our survey was derived from an *a priori* set of hypotheses sketched by Axhausen (2007), that cover discussion in its final form (Axhausen, 2008), initial qualitative work (Ohnmacht and Axhausen 2005; Larsen at al. 2006), related quantitative work on mobility biographies (Beige and Axhausen 2006; Ohnmacht et al. 2008) and a substantial pre-test (Frei 2007). The survey instruments address the following elements: first, basic socio-demographics of the respondent today, second, various 'mobilities' of the respondents, such as the mobility biography of residential and employment moves over a lifetime, including information about income levels, mobility tool ownership and main modes of transport to work and third, four name-generators and a name interpreter, that include exact home location of the respondents' social contacts and the frequency of their interactions by four modes: face-to-face, phone, email and texting (short-message-service – SMS). The survey tries to shed new light on current social practices in building and maintaining social networks. The survey overcomes the limitations of previous sources about the spatial patterns of social interactions, which were, by definition, partial to a particular mode: travel and activity diaries (face-to-face contacts), telecommunication diaries (phone plus a subset of the electronic channels: email, SMS, chat). It is also more comprehensive than the small number of previous surveys that covered multiple modes, but did not identify social network members involved (combined travel and (tele)communication surveys).

Data Collection

Data collection was conducted from December 2005 to December 2006. During a pre-test (Axhausen et al. 2006 and Frei, 2007), three different survey methodologies (self-completion, face-to-face, mixed face-to-face and self-completion) were tested to identify a survey format that could minimise missing values (due to fatigued interviewees) and reduced recall problems for retrospective survey items. The mixed method was the most effective approach with an acceptable cost per response (110 CHF/usable response). For the survey, 4,200 Zurich residents with available addresses and telephone numbers were chosen randomly. Following an announcement letter, the subjects were contacted on different days of the week and times of day, and then recruited during the telephone interview, including arranging appointments for the face-to-face interviews. With the reminder notice for the interview, respondents received the written form allowing them to raise questions during the upcoming interview. The written part consists of a person form and a form with mobility biographical questions about relocations, former and current job locations, usage of mobility tools, important life events and memberships in groups that meet periodically (see Beige 2006 for detailed information about mobility biographies). The one hour face-to-face interview covered the social contact questionnaire, but was also used to detect and address respondent difficulties and to establish rapport with the respondent.

The interviewers reached 2,714 (64.4 per cent) subjects by phone within five attempts. Of these, they could recruit 332 people, of whom 307 (10.7 per cent) were interviewed and completed the questionnaire. (For further details see Frei 2007.) The interviewees received no incentive. Due to the high response burden, the response rate is acceptable and within expectations. Furthermore, the response rate is satisfactory, given that the questionnaire was comprehensive.

Table 5.1 shows the socio-demographic characteristics of respondents in comparison with the general Zurich population, as observed in the Swiss Microcensus Travel 2005 (Swiss Federal Statistical Office and Swiss Federal Office for Spatial Development 2007) and the Swiss Census 2000 (Swiss Federal Statistical Office 2000). The income information is not directly comparable because the Microcensus measures household income, while this study is person-based. The comparison shows that the Zurich population is somewhat older, slightly better educated and has a higher share of public transport season tickets. Obviously, there is a slight bias towards a better-educated and public transport-oriented urban milieu. Still, an overall reweighing of the data seems unnecessary, given the relatively small deviations.

Table 5.1 Socio-demographic comparison between the characteristics of the Zurich respondents and the Zurich population[*]

Variable	Survey mean	Population mean	Difference
Age	50.76	46.76	+8.5%

Variable	Survey share	Population share	Difference
Male	43.6%	47.9%	−4.3%
Education			
NA	5.2%	12.5%	+7.3%
Obligatory schooling	8.0%	19.2%	−11.2%
Vocational training	31.8%	31.3%	−0.5%
High school diploma	8.3%	9.2%	−0.9%
Further technical training	20.8%	10.7%	+10.1%
University degree	26.0%	17.1%	+8.9%
Car available			
Always	44.6%	42.8%	+1.8%
Frequently and rarely	17.0%	18.4%	+1.4%
Public transport tickets			
50% discount card	49.5%	37.9%	+11.6%
National season (GA)	24.6%	14.2%	+10.4%
Regional season	13.8%	18.7%	−4.9%
Personal income (SFr/month)			
NA	12.8%		
0–1999	13.8%		
2000–5999	46.4%		
6000+	27.0%		

[*] As observed in the *Swiss Census 2000* (Bundesamt für Statistik, 2000) for age, gender and education and *Microcensus Travel 2005* (Bundesamt für Statistik und Bundesamt für Raumentwicklung, 2007) for the rest.

Surveying the Size of Social Contacts

To get a clear picture when surveying social network size, it is necessary to review the methodology of information gathering from respondents. We focus on egocentric (personal) networks, meaning that we use the respondent (alter) as the core of the social network and specifically capture his or her social contacts (alteri). To survey egocentric networks, name-generators and name-interpreters are used. The name generator specifies the type of relationship that the survey wants the respondent, the ego, to list. Often, researchers set an arbitrary maximum number of contacts to be listed (see, for example, Diaz-Bone 1997). The name-generator defines the egocentric network and is the basis for further analysis. The name-interpreters then pose further questions to detail the description of the contact, alter, e.g. socio-demographic data or characteristics of the relationship.

Most egocentric networks surveys today use name-generators appropriate to a stimulus. With a stimulus, a certain kind of activity is suggested, e.g. discussing important matters, for which the interviewee names alteri. For our research goal – measuring size and structure of social network geographies against the background of a concrete activity space – it is important to survey those alters with co-present intermittent visits. We used an adapted, appropriate set of name-generators as stimuli, and each respondent was handed two lists with two different name-generators. The first name generator was a variation of Burt's and Fischer's survey instrument (Burt 1984 and Fischer 1982) that asked for contacts with whom the respondents 'discuss important problems, with whom you stay in regular contact or who you can ask for help'. The second name generator asked for persons with whom the respondents spend leisure time. This generator targets weaker ties with the rationale that leisure travel makes up the largest share of long distance travel. The name-interpreter asked (for all of the named contacts) how they met, how long the relationship has existed, frequency of contacts by different modes (face-to-face, telephone, email and SMS – short message service via mobile phone), where they met the last time and the contact's place of residence. The origin of the acquaintance was categorised as family, subdivided in first degree, relatives or partner, work-related, education or partner or 'other'. Attempts were made to specify frequency of contacts as accurately as possible; e.g. every week, or twice a year. The contact's place of residence was clarified as much as possible with postal code, municipality, street and house number.

A first point of discussion concerns answer validation. Respondents were able to name a total of 17 relationships with the first name-generator and 32 with the second, producing a total of 49 alteri. In fact, the lists could have been extended if necessary. The range of named relationships is 1 to 49. The maximum number of reported relationships was reached once and the mean was 12.35 relationships. Compared to the possible number of 49 relationships, the exhaustion rate is 25.2 per cent, indicating that respondents had sufficient possibilities to cite relationships. Figure 5.1a indicates distribution of the number of relationships. The distribution is left skewed and has a variance of 73.0. The share of important relationships is 52 per cent and drops, as expected, with an increasing number of reported relationships (Figure 5.1b).

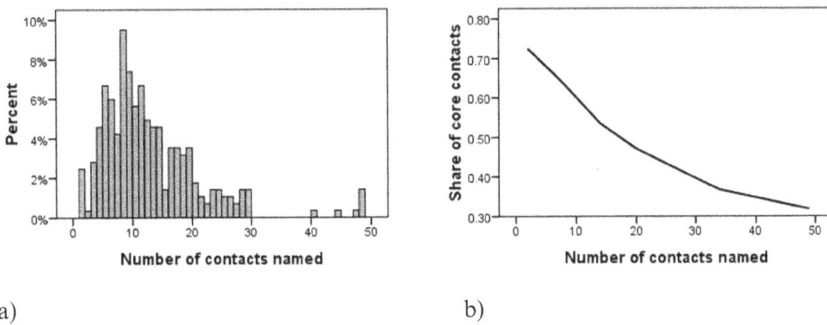

a)   b)

Figure 5.1   Distribution of the number of relationships

Measuring Spatial Dispersion of Social Network Geography

We will now describe the approach taken to capture spatial dispersion of social network geography to model respondents' differences in the empirical part of the chapter. Both biologists – and more recently, transport planners – have had to address the question of measuring spatial distributions in their analysis of daily activity spaces. They proposed parametric, semi-parametric, and non-parametric approaches to measure the size of the activity spaces (see Schönfelder 2006 for a review). The most popular (but also problematic) approach is to calculate the size of the confidence ellipse, i.e. the two-dimensional generalisation of the confidence interval (see Figure 5.2 for an example).

It is a parametric approach; the form of the approximation is fixed and the normal distribution of locations is assumed. The symmetry of the confidence ellipse often produces cases where half the area covered by the ellipse is empty of locations and therefore too big. Rai et al. (2007) suggest other geometries which overcome this problem, but incur substantial computational costs. They also found that the complex geometries correlate highly with the confidence ellipse.

The easiest way to capture the geography would be summing up the distances of egocentric social network ties. Unfortunately, this approach fails because the distance variable alone ignores the contacts' spatial distribution pattern, e.g. the agglomeration of contacts, which cannot be measured just by distance and its distribution patterns. The frequency-weighted sum of contact distances correlates very weakly with the 95 per cent confidence ellipses.

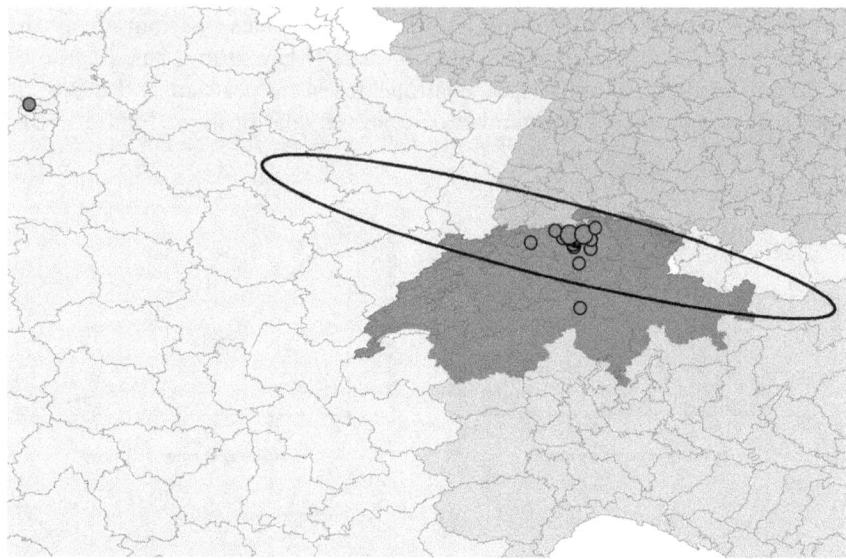

Figure 5.2   Example social geography measured by confidence ellipse

Note: The respondent is female, 61 years old, a homemaker and has moved four times in the last 46 years. The circles tag the home locations of the acquaintances and the sizes are proportional to the number of face-to-face contacts.

Empirical Results: Size and Spatial Dispersion

The following section highlights several major empirical findings on mobilities and social networks in detecting patterns of inequality. We begin with a size analysis of the egocentric network. Additionally, we focus on the spatial dispersion of social network geographies measured as confidence ellipses, followed by an analysis of differences between respondents.

Social network size   The number of social relationships reported is compared in Table 5.2, differentiated by respondents' socio-demographic characteristics. Age makes a large difference, with younger people cultivating more relationships than older ones. The share of important relationships appears to increase with age, but decreases slightly in the oldest age class. Gender seems to have no influence on the numbers of relationships. A higher level of education seems to increase the number of relationships, but the differences are small and there is no clear trend visible. The share of important relationships is slightly higher for persons with higher levels of education. There seems to be no income dependence.

In further analysis, it is necessary to determine the probability distribution that represents the data set to capture a wide range in the number of social relationships. Figure 5.1 shows that the data follows a left skewed bell-shaped

Table 5.2 Number of social relationships by socio-demographic characteristics

Variable	Median		Mean		St. dev.	
Category	Strong ties	All	Strong ties	All	Strong ties	All
Age						
Up to 30	6	12.5	7.1	15.1	3.6	9.4
30 to 40	5	14.0	6.8	14.0	3.3	5.4
40 to 60	5	10.0	6.7	12.4	4.4	9.2
60 and older	5	9.0	5.6	10.6	3.6	8.0
Sex						
Female	5	11.0	6.4	12.6	4.0	7.8
Male	5	11.0	6.2	12.3	3.6	8.7
Education						
N.A.	5	11.0	4.9	10.8	1.6	4.4
Obligatory schooling	5	8.0	5.5	12.5	2.7	11.9
Vocational training	5	11.0	6.1	11.6	3.9	8.0
High school diploma	5	12.0	6.5	14.1	4.0	9.0
Further technical training	5	8.5	6.1	10.6	3.9	6.4
University degree	5	13.0	7.2	14.5	3.4	8.2
Income (SFr/month)						
N.A.	5	10.0	6.0	11.3	3.3	6.6
0–1999	5	12.5	5.9	12.5	3.1	6.1
2000–5999	5	11.0	6.6	13.1	3.9	8.7
6000+	5	11.0	6.3	12.1	4.3	9.4
All	5	11.0	6.0	12.0	4.0	8.0

curve. To deal with the skew of number of relationships, a negative binomial distribution is used to represent this shape. For modelling the number of relationships, six persons reporting very high numbers (above 30 relationships) were removed as potential outliers.

Socio-demographic, travel-related, biographical and survey-specific dummy variables are employed to explain the number of social contacts using a negative binominal regression. After removing variables that correlate highly with each other (limit = 0.5; e.g. working status and place of work), variables with a significance level lower then 0.05 were removed stepwise. The parameter estimates are reported in Table 5.3.

Table 5.3 Parameter estimates for the negative binominal regression of the number of contacts

Variable	Mean	St. dev.	Beta	b/St. err	Sign.
Constant			3.10	10.17	0.00
Age (years)	53.3	19.2	–0.04	–3.11	0.00
Age^2/1000 (years^2/1000)	3.2	2.1	0.35	2.81	0.01
Annual or monthly public transport ticket (yes)	0.9	0.9	0.24	2.04	0.04
Number of relocations (n)	5.9	3.1	0.04	3.02	0.00
University degree (yes)	0.2	0.4	0.18	1.92	0.05
Part-time employed (yes)	0.2	0.4	–0.26	–2.32	0.02
Retiree (yes)	0.3	0.5	–0.30	–2.00	0.05
Children in the household < 18 y (yes)	0.3	0.4	0.18	2.31	0.02
N			300		
Adjusted R^2			0.13		

The goodness-of-fit is, as expected from the descriptive statistics, rather low ($R^2 = 0.13$), but the F-statistic is significant. Furthermore, the results show that age of the respondents shows a U-shaped influence. Younger people maintain many contacts and then the number declines with increasing age, whereas every additional year causes a lower decrease of the number of social relationships. Ownership of public annual or monthly transport ticket has a positive influence on the number of social relationships. Maintaining a larger social network seems to be influenced by ownership of mobility tools, but only an annual or monthly subscription to public transport tickets was highly significant. The number of relocations influences the number of relationships. However, the positive influence of even a number of relocations indicates that people keep their important friendships after moving, even over distance. By building up a social network at the new location and keeping in touch with 'old friends', numbers of social contacts increase. A higher education, at least a university degree, leads to a larger number of social contacts. But there is no clear trend visible between the number of social contacts and education. Working status has a strong influence on the number of social relationships. In 27.3 per cent of cases, the origin of the acquaintance is work (41.0 per cent friends, 25.9 per cent family, 4.9 per cent partner and 0.9 per cent others), making it the second most frequent original context. The big influence of work status is not surprising. Part-time employees and retired people have fewer social relationships than full-time employees and equivalents, e.g. students. It is noteworthy that children in the household have a positive influence on the number of social contacts. One might expect that the additional workload for parents would decrease the number of social relationships, but children facilitate possible new contacts: e.g. other parents with small children, parent-teacher conferences etc., which outweigh the first effect.

Spatial Dispersion – The Social Network Geography

To analyse spatial spread, we apply multivariate models. In a model of the 95 per cent confidence logarithm ellipses as a dependent variable, the values are all nonnegative, with 33 zero values in a total of 276 observations.[3] Conventional regression-methods, as the ordinary least square method, are not adequate for such censored values (Greene 2000).

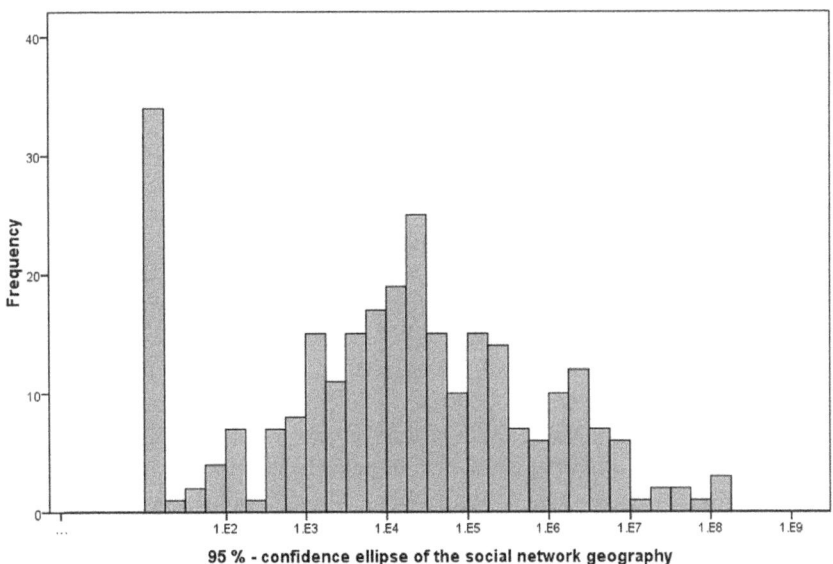

Figure 5.3   Distribution of the social network geometries measured as 95% confidence ellipses (km^2)

Note: Social network geographies of less than 10 km^2 were coded as 0.

---

3   The smaller sample comes from respondents with three or fewer distinct geocodes for their contacts, for which no ellipses can be calculated.

The size distribution of the 95 per cent confidence ellipses seems to follow a log-normal distribution (Figure 5.3), if we ignore the third of the respondents who have a local set of contacts. The fit statistics for a lognormal distribution are good and do not reject this distribution at the 0.05-level (chi-square, Kolmogorov-Smirnov and Anderson-Darling Test) (estimated with ExpertFit – Version 7.00 see Averill M. Law and Associates 2006). In comparison, other distributions – such as the Weibull, Gamma, log-Logistic and several others – performed less well.

The Tobit model is able to differentiate between limit- and non-limit-observations. This model assumes that the limit outcome is determined by the level of the non-limit outcome. To test this assumption, a different model, also appropriate for the data, can be compared to the Tobit model. This is Cragg's Model for Censored Data (Cragg 1971). It is used when the assumption of the Tobit model (where the non-limit outcome is determined apart from the level of the non-limit outcome) is not true. Cragg's Model is a combination of the Probit model (for y = 0) and the truncated regression (for y>0). The zeroes in our data have their origin in two different problems. First, only 44 per cent of the geocodes have street address accuracy; the rest have only zip-level accuracy, leading to just one geocode for several contacts. Second, the confidence ellipse needs at least three spatially distinct locations to be calculated. The origin of the zeros in the data leads to the assumption that the non-limit outcome is determined by the same level as the limit outcome, which is shown through the Probit and Tobit results (Table 5.4).

The Tobit model was calculated after removing variables that correlate highly with each other (limit = 0.5). Variables with a significance level lower than 0.05 were removed stepwise. The parameter estimates are reported in Table 5.4.

Analysis of the Tobit results shows that there are different factors influencing social network geographies. The first group consists of socio-demographic variables. The model results indicate that young people with high education and low or middle income tend to maintain a more spatially distributed social network. The influence of age and education is similar to their influence on the numbers of relationships. The influence of income is unexpected, as a spatially more distributed social network is expensive to maintain. One interpretation of the negative influence could be that a higher income is often linked to a higher workload and more responsibility, leading to a higher time value for these persons. Since travel costs have decreased (see, for example, Fröhlich et al. 2006), time costs now seem to exceed the financial costs of travelling. Car ownership has a positive influence on the size of social network geographies; even if ownership does not contribute to maintenance of contacts over distance (see above), it is an indication of mobility. Number of relationships has an influence, as mentioned above, because it is correlated with the share of non-core contacts. It is now possible to spatially maintain more widely distributed networks of weaker ties with less frequent face-to-

Table 5.4 Parameter estimates for the Tobit regression of the logarithm of the size of the 95% confidence ellipses and the associated Probit model of the Cragg approach

			Tobit model		Probit model	
Variable	Mean	St.dev.	Beta	Sign.	Beta	Sign.
Constant	-	-	9.92	0.00	2.45	0.03
Age (years)	53.4	19.3	-0.29	0.00	-0.11	0.01
Age^2/1000 (years^2/1000)	3.2	2.1	2.94	0.00	1.10	0.01
Car ownership (yes)	0.5	0.5	1.60	0.01	0.19	0.37
Number of relationships	12.5	8.4	0.20	0.00	0.09	0.00
Education/workplace changes	3.3	2.4	0.28	0.02	0.06	0.28
Further technical training (yes)	0.2	0.4	2.48	0.00	0.58	0.04
University degree (yes)	0.2	0.4	2.61	0.00	0.40	0.16
Income >6000 CHF (yes)	0.3	0.4	-1.64	0.03	-0.28	0.24
N		286			241	
Goodness-of-fit		Adjusted R^2 = 0.25		Chi^2 (8 df) = 47.31		

face contacts by using telecommunication contacts (see Axhausen 2007 for details). The number of education or workplace moves has a biographical influence on the social network geographies. Apparently, being less anchored in space and being professionally flexible have a positive influence on the size of the social network geographies. Surprisingly, spatial distribution of education and workplace changes, measured by their confidence ellipses, has no significant influence on them. Overall, the model explains 25 per cent of the variance of social network geographies. The parameters of the Probit model exactly follow the parameters of the Tobit model (See Table 5.4). The resulting predictions are 100 per cent correct for the 1s (y>0) and 22.5 per cent correct for the 0s, resulting in overall 89.2 per cent correct values. As the parameters of the Probit model show, limit outcome is determined by the level of non-limit outcome, so the estimates of the truncated model for non-limit observations are omitted.

Conclusion

Our primary goal for this chapter was to present insights into the conceptual background of the issue of mobilities and social networks. We first discussed relevance for social sciences and transport studies by presenting the pertinent strands of discussion in recent literature. We then explained methodological challenges in data gathering to examine personal mobilities and social networks. Finally, we presented several main empirical findings from a representative Zurich survey. In the following, we briefly conclude by discussing the empirical result.

Social network geographies indicate geographical patterns of personal relationships, especially how spatially distributed they are. We found that size and spatial dispersion of social network geographies differ according to various stratification dimensions, which in turn are related to various mobilities and inequality patterns.

In general, the analysis shows that the distribution of network geographies is very wide – from local ties to international ties – and a remarkable share of intercontinental ties. We examined the effects of residential change on spatial dispersion of an egocentric network (see also Ohnmacht et al. 2008). An explanatory factor is mobility biography. We see evidence that the more 'events' producing change occur – here especially measured by the number of relocations – the more spatially dispersed the network geography is. This finding can be linked to the forms of life flexibility discussed, for instance, by Kesselring (2008, 78):

> Within the mobile risk society people are self-responsible for the roads (metaphorically speaking: the authors) and trajectories they choose during their life course.

According to this sociological diagnosis and empirical findings, subjects might free themselves from both local obligations and responsibilities, falling back on local contacts. The consequence is a movement toward more flexibility in organising ones life and disappearance of the traditional 'normal biography'. This results in spatial diversity and social differentiation. Beck, for instance, talks about 'Issue-Communities' which are not necessarily locally integrated, but are instead based on reciprocal interests, such as leisure time activities, etc. (Beck 1992). Thus, for certain groups, social life becomes more fluid and dispersed, as well as long- distance (Lash and Urry 1994).

In summary, new questions evolve. What is the impact of having widespread social relations and how can one maintain it? What other factors cause social networks to become more spatially dispersed over time, influencing our travel behaviour? Why is it suddenly necessary for many people to travel long distances to meet with friends, relatives and partners? Which 'mobility tools' such as cars, public transport, bicycles and the new age of low-cost airlines, (and to what extent) are necessary to meet them in person and to maintain the relationship? Long distance ties figure in more than half of egocentric networks. This fact is reflected in long distance travel statistics where the highest share involves visiting friends and relatives (e.g. Federal Statistical Office and Swiss Federal Office for Spatial Development 2007). It should be noted in further research and modelling that social network geographies have a certain structure at a certain size. These first results in analysing social network geographies patterns show that the ego's characteristics, (mainly socio-demographics and mobility biography events), can to a certain extend explain them.

References

Averill M. Law and Associates (2006), *ExpertFit Version 7* (Tuscon: Unifit).
Axhausen, K.W. (2007), 'Activity Spaces, Biographies, Social Networks and their Welfare Gains and Externalities: Some Hypotheses and Empirical Results', *Mobilities* 2:1, 15–36.
Axhausen, K.W. (2008), 'Social Networks, Mobility Biographies and Travel: The Survey Challenges' *Environment and Planning B* 35:6, 981–96.
Axhausen, K.W., Frei, A. and Ohnmacht, T. (2006), 'Networks, Biographies and Travel: First Empirical and Methodological Results', paper presented at the 11th International Conference on Travel Behaviour Research, August, Kyoto.
Axhausen, K.W., Löchl, M., Schlich, R., Buhl, T. and Widmer P. (2007), 'Fatigue in Long Duration Surveys', *Transportation* 34:2, 143–60.
Axhausen, K.W., Dolci, C., Frölich, P., Scherer, M. and Carioso, A. (2008), 'Constructing Timescaled Maps: Switzerland from 1950 to 2000', *Transport Review* 28:3, 391–413.
Beck, U. (1992), *Risk Society: Towards a New Modernity* (London: Sage Publications).
Beck, U. (2008), 'Mobility and the Cosmopolitan Perspective', in Canzler, W., Kaufmann, V. and Kesselring, S. (eds), *Tracing Mobilities: Towards a Cosmopolitan Perspective* (Aldershot: Ashgate), 25–36.
Beck-Gernsheim, E. (2007), 'Transnational Lives, transnational Marriages: A Review of the Evidence from Migrant Communities in Europe', *Global Networks* 7:6, 271–88.
Beige, S. and Axhausen, K.W. (2006), 'Long-term Mobility Decisions during the Life Course: Experiences with a Retrospective Survey', paper presented at the 11th International Conference on Travel Behaviour Research, August, Kyoto.
Blake, R., Rhead, C., Wedge, B. and Mouton J.S. (1956), 'Housing Architecture and Social Interaction', *Sociometry* 19:2, 133–9.

Burt, R.S. (1984), 'Network Items should be included in the General Social Survey', *Social Networks* 6:4, 293–339.
Butts, C.T. (2003), 'Predictability of Large-scale Spatially Embedded Networks', in Breiger, R., Carley, K.M. and Pattison, P. (eds), *Dynamic Social Network Modeling and Analysis: Workshop Summary and Papers* (Washington, DC: National Academies Press), 313–23.
Caplow, T. and Foreman, R. (1950), 'Neighborhood Interaction in a homegenous Community', *American Sociological Review* 15:3, 357–66.
Carrasco, J. and Miller, E.J. (2006), 'Exploring the Propensity to perform Social Activies: A social Network Approach', *Transportation*, 33:5, 463–80.
Carrasco, J., Hogan, B., Wellman, B. and Miller, E.J. (2008), 'Collecting Social Network Data to Study Social Activity-Travel Behavior: An Egocentric Approach', *Environment and Planning B* 35:6, 961–80.
Cass, N., Shove, E. and Urry, J. (2005), 'Social Exclusion, Mobility and Access', *Sociological Review* 53:3, 539–55.
Cragg, J. (1971), 'Some Statistical Models for limited dependent Variables with Application on the Demand for Durable Goods', *Econometrica* 39:5, 829–44.
Diaz-Bone, R. (1997), *Ego-zentrierte Netzwerkanalyse und familiale Beziehungssysteme* (Wiesbaden: Deutscher Universitäts-Verlag).
Federal Statistical Office and Swiss Federal Office for Spatial Development (2007), *Mikrozensus zum Verkehrsverhalten 2005* (Bern: Bundesamt für Statistik).
Fischer, C.S. (1982), 'What Do We Mean by 'Friend?' An Inductive Study', *Social Network* 3:4, 287–306.
Frei, A. (2007), 'Feldbericht zu den sozialen Netzwerkbefragungen', Working Paper, *Arbeitsberichte Verkehrs- und Raumplanung*, 440, IVT, ETH Zürich, Zürich.
Fröhlich, P., Tschopp, M. and Axhausen, K.W. (2005), 'Entwicklung der Erreichbarkeit der Schweizer Gemeinden: 1950 bis 2000', *Raumforschung und Raumordnung* 63:6, 385–99.
Götz, K. (2007), *Freizeit-Mobilität im Altag oder Disponible Zeit, Auszeit, Eigenzeit – warum wir in der Freizeit raus müssen* (Berlin: Duncker und Humblot).
Greene, W.H.H. (2002), *Econometric Analysis* (New York: Prentice Hall).
Grieco, M., Turner, J. and Hine, J.P. (2001), 'Transport, Employment and Social Exclusion', *Local Work*, vol. 26 (Manchester: Centre for Local Economic Strategies).
Handy, S.L. and Niemeier, D.A. (1997), 'Measuring Accessibility: An Exploration of Issue and Alternatives', *Environment and Planning A* 29:7 1175–94.
Hine, J.P. and Grieco, M. (2003), 'Scatters and Clusters in Time and Space: Implications for delivering integrated and inclusive Transport', *Transport Policy* 10:4, 299–306.

Holmes, M. (2004), 'An Equal Distance? Individualisation, Gender and Intimacy in Distance Relationships', *Sociological Review* 52:2, 180–200.
Jones, P.M., Dix, M.C., Clarke, M.I. and Heggie, I.G. (1983), *Understanding Travel Behaviour* (Aldershot: Gower).
Kaufmann, V., Bergman, M.M. and Joye, D. (2004), 'Motility: Mobility as Capital', *International Journal of Urban and Regional Research* 28:4, 745–65.
Kesselring, S. (2008), 'The Mobile Risk Society', in Canzler, W., Kaufmann, V. and Kesselring, S. (eds), *Tracing Mobilities: Towards a Cosmopolitan Perspective* (Aldershot: Ashgate).
Larsen, J., Urry, J. and Axhausen, K.W. (2006), *Mobilities, Networks, Geographies* (Aldershot: Ashgate).
Latane, B., Liu, J.H., Nowak, A., Bonevento, M. and Zheng, L. (1995), 'Distance Matters: Physical Space and Social Impact', *Personality and Social Psychology Bulletin* 21:8, 795-804.
Löw, M. (2001), *Raumsoziologie* (Frankfurt/Main: Suhrkamp).
Löw, M. (2005), 'The Constitution of Space. The Double Existence of Space as Structural Ordering and Performative Act', lecture at Paris I/Sorbonne 14 March 2005, <http://www.raumsoziologie.de> (homepage), accessed 21 May 2008.
Marsden, P.V. (2005), 'Recent Developments in Network Measurement', in Carrington, P.J., Scott, J. and Wasserman, S. (eds), *Models and Methods in Social Network Analysis* (New York: Cambridge University Press), 8–30.
Ohnmacht, T. and Axhausen, K.W. (2005), "Wenn es billiger ist als die Bahn – na ja, warum nicht?': Qualitative Auswertung zu Mobilitätsbiographien, Mobilitätswerkzeugen und sozialen Netzen', Working Paper, *Verkehrs- und Raumplanung*, 313, IVT, ETH Zürich, Zürich.
Ohnmacht, T., Frei, A. and Axhausen, K.W. (2008), 'Mobilitätsbiografie und Netzwerkgeografie: Wessen soziales Netzwerk ist räumlich dispers?', *Swiss Journal of Sociology* 31:1, 131–64.
Ortuzar, J.D. and Willumsen, L.G. (2001), *Modelling Transport* (Chichester: Wiley).
Prillwitz, J., Harms, S. and Lanzendorf, M. (2006), 'Impact of Life Course Events on Car Ownership', *Transportation Research Records* 1985, 71–7.
Rai, R.K., Balmer, M., Rieser, M., Vaze, V.S., Schönfelder S. and Axhausen, K.W. (2007), 'Capturing Human Activity Spaces: New Geometries', *Transportation Research Record* 2021, 70–80.
Scheiner, J. (2007), 'Mobility Biographies: Elements of a Biographical Theory of Travel Demand', *Erdkunde* 61:2, 161–73.
Schönfelder, S. (2006), 'Urban Rythms: Modelling the Rythms of Individual Travel Behaviour', dissertation, ETH Zürich, Zürich.
Schönfelder, S. and Axhausen, K.W. (2003), 'Activity Spaces: Measures of Social Exclusion?', *Transport Policy* 10:4, 273-86.

Schubert, S. (1994), 'Zur Bedeutung von räumlichen Entfernungen und sozialen Beziehungen für Hilfe im Alter', *Geographische Zeitschrift* 82:4, 226–38.
Sheller, M. and Urry, J. (2006), 'The New Mobilities Paradigm', *Environment and Planning A* 38:2, 207–26.
Shields, R. (1991), *Places on the Margin. Alternative Geographies of Modernity* (London: Routledge).
Simmel, G. (1908), *Soziologie. Untersuchungen über die Formen der Vergesellschaftung* (Berlin: Duncker und Humblot Verlag).
Snow, D.A., Leahy, P. and Schwab, W. A. (1981), 'Social Interaction in a Heterogeneous Apartment: An Investigation of the Effects of Environment upon Behaviour', *Sociological Focus* 14:4, 309–19.
Sommer, R. (1996), *Personal Space* (Englewood Cliffs, NJ: Prentice Hall).
Swiss Federal Statistical Office (2000), *Eidgenössische Volkszählung (VZ) 2000* (Bern: Bundesamt für Statistik).
Swiss Federal Statistical Office and Swiss Federal Office for Spatial Development (2007), *Mikrozensus zum Verkehrsverhalten 2005* (Neuchâtel and Berne: Swiss Federal Statistical Office and Swiss Federal Office for Spatial Development).
Tönnies, F. (1991), *Gemeinschaft und Gesellschaft: Grundbegriffe der reinen Soziologie* (Darmstadt: Wissenschaftliche Buchgesellschaft).
Urry, J. (2008a), *Mobilities* (Oxford: Blackwell).
Urry, J. (2008b), 'Moving on the Mobility Turn', in Canzler, W., Kaufmann, V. and Kesselring, S. (eds), *Tracing Mobilities: Towards a Cosmopolitan Perspective* (Aldershot: Ashgate), 13–24.
Wasserman, S. and Faust, K. (1994), *Social Network Analysis: Methods and Applications* (New York: Cambridge University Press).
Wellman, B. (1996), 'Are Personal Communities Local? A Dumptarian Reconsideration', *Social Networks* 18:4, 347–54.
Wellman, B. (1979), 'The Community Question: The Intimate Networks of East Yorkers', *The American Journal of Sociology* 84:5, 1201–31.
Wittel, A. (2001), 'Towards a Network Sociality', *Theory, Culture and Society* 18:1, 51–76.

Chapter 6
Social Integration Faced with Commuting: More Widespread and Less Dense Support Networks

Gil Viry, Vincent Kaufmann and Eric D. Widmer

Introduction

Boltanski and Chiapello (1999) remind us that today, the ability to move is essential not only to peoples' careers, but also to their social integration in general. Mobility has become a central aspect of social integration, notably by contributing to transformation of the modalities of relational embeddedness and the space in which these are implemented. Indeed, the speed potentials afforded by transportation and communication systems allow people to build more distant social ties. In a context where spheres of activity within a single day have both greatly increased in number and grown in distance, mobility potentials may be used as a resource to ward off those spatial and temporal incompatibilities that actors must contend with. In highly advanced societies that have seen an increase in the ways that people can travel through time and space (Urry 2000; Kaufmann 2002), mobility is a value that carries its own differentiations. Its effective use allows a person to acquire social status, whereas neglecting mobility may lead to loss of status. Therefore, this growing importance of spatial mobility contributes to the creation of new forms of inequality. Not having a car (Dupuy 1999), living in a residential area with poor access to public transportation (Cass et al. 2005; Jemelin et al. 2007), and weak temporal or organisational resources to handle projects that require travel (Kaufmann et al. 2005; Le Breton 2005) may jeopardise the social and professional integration of disadvantaged portions of a population.

Contrary to the urban sociology of the 1930s, which saw in the explosion of big cities a risk of anonymisation and social disaffiliation in metropolitan contexts, sociology has since emphasized the plurality of social integration forms (Wellman 1988), opposing the thesis of disaffiliation. In this same idea, some authors (Offner and Pumain 1996; Kesselring 2006a, 2006b; Frei et al., Chapter 5, this volume; Ohnmacht et al. 2008) have suggested that social links are built less in the proximity and the public space and more in relatedness and distance relationships. The development of commuting in the 1970s, which relaxed the spatial dependence between the workplace and the residence, pertains to this transformation of social anchorings through spatial mobility. The increase of travel time budget and geographical distances is a challenge to the constitution of social ties, whose certain forms are mainly forged in habit and daily time. This chapter addresses these issues on the basis of new Swiss data by examining the spatiality of social integration in a commuting context. From the concept of social capital, it asserts that, if commuting weakens local relationships, it reconstructs more decentralized integration forms, presenting other relational constraints and opportunities. Social inequalities resulting from this new geography of social integration are also discussed based on the concept of *motility*.

The Transformations of the Spatiality of Social Integration

The metropolisation process operating in Switzerland for about 15 years is generic and singular. It is generic because, like most European countries, the largest Swiss agglomerations – Zurich, Geneva and Basle – concentrate the bulk of job creation, leading to an increase in commuting to these destinations. For example, commuter traffic between the major Swiss cities (Zurich, Basle, Geneva, Berne and Lausanne) has doubled or tripled every decade (Frick 2004). The metropolisation process is also singular because the metropolisation of large urban centres manifests itself by new dynamics, directly affecting those centres' hinterlands. Indeed, bi-residentiality and long distance commuting between urban centres and rural areas are quickly developing in Switzerland, benefiting peripheral regions while increasing travel in terms of flow and budgeting time.

Important transformations of the relations with space and, more particularly, of the spatiality of social integration are behind these trends. More than half of the working Swiss population does not work in their municipality of residence. Therefore, the residence neighbourhood is not necessarily the theatre of daily life any more. The change occurred very quickly: about 50 years ago, the Swiss population was mostly non-motorised, so activities and social relations were centred on home neighbourhoods. With the development of commuting and the emergence of long-distance commuting and bi-residentiality, social integration is no longer limited to the proximity of a residence.

One result of this is that the classical distinction of daily mobility, related to travel centred on the daily living environment and residential mobility, related to a social uprooting has partly lost its relevance. It is more common to have a daily life that takes place in areas that are dozens of kilometres apart from each other, with habits and routines in each of those areas. A second result is the development of *poly-places*, anchoring forms that are built around attachments to places and/or around social relationships. Confronted with mobility demands, more individuals are forced to develop and maintain social anchorings in different places, sometimes far apart from each other. In this configuration of multiplication and spatial expansion of relational anchorings, individuals should access to various means of transportation (e.g., cars, trains, planes) and various forms of mobility (e.g., physical, virtual, phone) to become socially integrated. According to the chosen strategies and resources that they command, individuals may strive to connect these different relational anchorings with each other or to maintain them unconnected (Kennedy 2004, 2005), possibly leading to the production of *spatial multi-belonging forms*.

Social Capital and Commuting

A frequent topic of urban sociology research since the findings of Chamboredon and Lemaire (1970), the link between commuting and social relationships is a central dimension of sociology related to social integration forms specific to modernity. Do social relationships change because of commuting? Though this question is well-known, it has not been the subject of systematic investigations in Switzerland. Contrary to the pessimistic hypotheses of the classics, which denounces the decisive weakening of social integration in metropolises, we suggest the hypothesis that commuting affects more the structure than the amount of interpersonal links a person has. In this perspective, we consider the concept of social capital.

The notion of social capital was used in sociology by Bourdieu (1980), as well as Coleman (1988) and Granovetter (1982, 2000). *Social capital* is classically defined as 'the set of current or potential resources stemming from the possession of a lasting network of more or less institutionalized relations' (Bourdieu 1980, 2, own translation). It is a set of relations specific to each individual, which can be considered as resources through the capability given to this individual to mobilise the people with whom he or she is connected.

The literature on the composition of social capital distinguishes two types of ties, strong ties and weak ties. Granovetter (1982) differentiates strong ties (i.e., durable, multiplex ties involving frequent interactions with a strong emotional implication) from weak ties (i.e., the acquaintances from diverse activity fields, like work or leisure). These two types of ties result in three types of social capital: capital based on strong ties (*binding social capital*), capital based on weak ties (*bridging social capital*) and capital that combines both types of ties (*binding-bridging social capital*) (Widmer 2006). Binding social capital corresponds to the *closed networks* of Coleman (1988), which are densely connected networks with a low centralisation. Most of the individuals, if not all, in closed networks are interlinked by significant ties. Relational constellations tend to be transitive. If an individual called Ego is linked to an Alter X and an Alter Y, it is likely that X is also linked to Y. Conversely, bridging social capital is associated to sparsely connected networks, characterised by a high centralisation and weak transitivity, leading some actors to benefit from a position of *compulsory intermediaries* between different network members (Burt 1992, 2002) (see Figure 6.1). If the combination of strong and weak ties corresponds to a level of social capital, it is not possible to rank in terms of quantity of social capital the binding and bridging types (Portes 1998; Wilson 1987; Granovetter 1982).

The two types of capital have particular advantages and disadvantages. On the one hand, binding social capital integrates the individual in a dense network of solidarity. On the other hand, it binds the person by strong social control. Bridging social capital provides the individual with more autonomy, but it puts the person in a position of relative weakness according to solidarity practices, which can only be expressed in an individual way because the network members are not linked to each other.

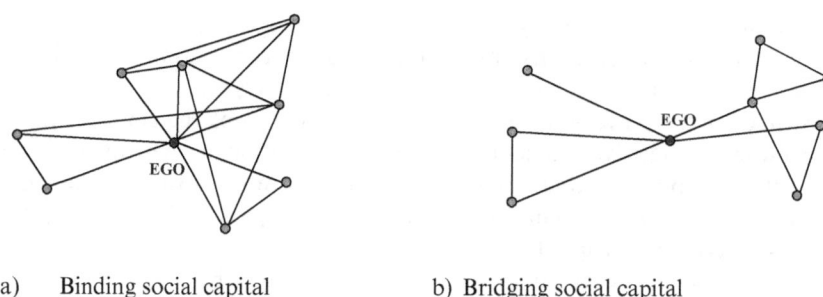

a)   Binding social capital   b) Bridging social capital

Figure 6.1   Illustration of binding and bridging social capital

From this point of view, we can make the hypothesis that commuters are more likely to develop bridging social capital than non-commuters. The distance between a place of residence and a workplace gives relational anchorings a particular configuration. Presumably less bound in neighbourhood relationships (Putnam 2000), commuters develop their interpersonal relationships in a broader spatial range, which does not necessarily weaken their networks, but does make them more spatially diverse and less connected. Relational anchoring in several places prevents commuters from putting in touch their significant others. For example, it is more difficult for commuters to benefit from network support if other network members would not or little support them each other. Commuting leads to a spread out and disconnected space relationship, although not necessarily a poorer relationship. The spatial multi-belonging corresponds to bridging social capital in its relational dimension and redefines the relationship of persons to space.

Study Data and Indicators

The 2005 MOSAiCH[1] survey included the Swiss portion of the yearly survey of the International Social Survey Program (ISSP). Data are composed of a representative sample of the population living in Switzerland 18 years old and older. Face-to-face interviews were conducted with 1,078 people on the basis of a standardised questionnaire. Respondents were asked about their social networks based on the following question: *From time to time, most people discuss important matters with other people. Looking back over the last six months, who are the people with whom you discussed matters important to you (work, family, politics, etc.)?* Respondents could mention a maximum of four persons (*significant others*). At the spatial level, each respondent was asked to identify for each network member (including themselves) the current municipality (*commune*) of residence, the municipality of residence at age 14 and the municipality of their current workplace. On the basis of that information and with the help of *routing* software designing the Swiss road network, road distances were computed. The geographical centres of the municipalities were used as coordinates. Three types of distance were extrapolated:

1. the distance between the residences of any two network members;
2. the distance between the current residence and the residence at the age of 14 (indicator of earlier residential mobility) of any network member;
3. the commuting distance of each active network member (indicator of spatial job-related mobility).[2]

From the first type of distance, we constructed two indicators of spatial expansion:

1. the mean distance between the respondent's residence and the one of each person mentioned by the respondent (*distance Ego-Alters*);
2. the mean distance between the Alters' residences (*distance Alter-Alter*).

These indicators enabled us to analyse the network spatial expansion according to two components. The first component was related to the relationship

---

[1] Sociological Measures and Observation of Attitudes in Switzerland. This study was funded by the Swiss National Science Foundation and was conducted by the Swiss Information and Data Archive Service for the Social Sciences (SIDOS).

[2] Only the Swiss municipalities were nominally stored in the database. When the person had lived, worked, or lived at the age of 14 outside the Swiss territory, the respective distance was then defined as missing values.

between the respondent (*Ego*) and each of the significant others (*Alters*). The second component was only related to the significant others. This last indicator had the advantage of eliminating any definitional dependencies in the analysis of relationships between respondent characteristics and network expansion measures.

The respondents were also asked to identify the person who gives *emotional, moral support* to others in the network.[3] Based on that information, the number of emotional support ties, mutual and not mutual, between the respondent and the significant others on the one hand, and between the significant others on the other hand, were computed. In order to construct support indicators that were independent of the network size, we also defined the *activation* of the support ties by the number of existing support ties divided by the number of potential support ties based on the number of persons mentioned by the respondent.[4] Table 6.1 presents the summary of the used variables for the data.

Commuting and Network Spatial Expansion

In order to study the effects of the respondents' commuting on their network spatial expansion, a linear regression analysis was carried out for each of the two indicators. The results of the analysis (Table 6.2) show that the effect is significant (level: 0.01). The more the respondent commutes, the farther away the persons mentioned in the network live from each other and the farther away the respondent lives from them. According to our regression model, for each increase of 10 km in the respondent's commuting distance, the significant others distance themselves from each other by an average of 2.35 km and the respondent distances him- or herself from them an average of 2.24 km.

In order to control the effect of different respondent's characteristics (see Table 6.2) on this outcome and to refine the analysis, a multiple regression analysis was conducted.

This analysis shows that the respondent's earlier residential mobility has the strongest influence on the physical distance between the respondent and his or her significant others (Table 6.3, left column). The next important factor is living alone. This last result stems largely from the fact that the respondents who lived alone did not mention any network members living with them, which caused an increase in the mean distance between them and their significant network members. Also important is the effect of commuting distance. Other

---

3   The questions were: *among these persons, who would give you some emotional, moral support at the time of everyday difficulties (for example, when you are a little bit depressed or following a hard day)? And which person or persons, you included, would give some emotional support to [first person mentioned]?* And so on.

4   This boils down to a calculation of density.

Table 6.1 Summary of the used scale variables

	Mean	St. dev.	Min. value	Max. value	No.
Distance Ego-Alters (in km)	19.7	35.4	1.45[a]	299.49	825
Distance Alter-Alter (in km)	27.4	37.5	0.92[a]	193.09	531[b]
Commuting distance of Ego (if active) (in km)	12.9	23.5	1.60[a]	241.00	675
Mean commuting distance of the Alters (in km)	13.1	17.6	1.09[a]	217.66	711
Distance between current residence and one at 14 y.o. of Ego (in km)	34.3	53.3	1.02[a]	354.00	882
Activation of the mutual support ties Ego-Alters (in ‰)	718.9	372.2	0	1000	919
Activation of the support ties given by Ego (in ‰)	842.0	311.8	0	1000	919
Activation of the support ties received by Ego (in ‰)	780.8	331.3	0	1000	919
Activation of the support ties between Alters (in ‰)	385.1	357.5	0	1000	620[b]

Notes

a   The minimum values of distance are not strictly equal to 0 because a value of 2 km was attributed in the situations where the departure and arrival municipality were identical, in order to take into account travel inside the municipality. Some rare inter-municipality distances are lower than 2 km.

b   The low number of cases is explained by the fact that about 40 per cent of the respondents mentioned fewer than two network members.

variables, such as sex, education, or the context of the respondent's residence do not influence considerably the physical distance between the respondent and the significant network members.

Concerning the physical distance between the significant others (Table 6.3, right column), it is again the respondent's earlier residential mobility that has the highest impact. Commuting has a slightly more important effect than in the case of the physical distance between a respondent and the significant others. Living alone, having a university degree and having a residence in an urban centre are also significantly associated, though more moderately, with a higher physical distance between significant network members. On the other hand, the gender of the respondent did not substantially affect the physical distance between the significant others.

Table 6.2 Frequency (in %) of the respondents' socio-demographic variables

	f		f		f
Sex		**Age (years)**		**Education**	
Female	53	18–34	22	Basic education	17
Male	47	35–50	30	Apprenticeship	41
		51–65	27	General education school	8
		66 +	21	High (professional) school	24
				University	10
	100		100		100
(N)	(1077)	(N)	(1078)	(N)	(1068)

	f		f
Family structure[a]		**Context of the residence**	
Person living alone	30	Peripheral commune	22
Couple living without children	27	Peri-urban commune	16
Couple living with children	29	Suburban commune	30
Person without cohabiting partner living with children	4	Small centre	11
Other family structures	10	Middle centre	12
		Big centre	9
	100		100
(N)	(1078)	(N)	(1078)

Note
a   The family structure has been defined on the basis of the cohabitation with the respondent. A couple is then defined by a partner cohabiting with the respondent.

Commuting and Emotional Support

Because the effects of commuting on the number of the emotional support ties were not found to be significant,[5] we focus on the link between commuting and activation of the support ties.

---

5   This result can be explained by a curvilinear effect observed in the relation between the network size and the respondent's commuting, which is not visible in a linear regression analysis.

Table 6.3 Regression analysis of network expansion on different variables related to the respondent (unstandardised regression coefficients)[a]

	Mean distance Ego-Alters		Mean distance Alter-Alter	
Commuting distance	0.22*** (0.15)	0.24*** (0.17)	0.24*** (0.16)	0.27*** (0.19)
Distance between current residence and the residence at 14 years old		0.23*** (0.33)		0.20*** (0.27)
Sex (female)		−1.53 (−0.02)		−1.86 (−0.02)
Age				
18–34 years old		0.63 (0.01)		0.18 (0.002)
35–50 years old		–		–
51–65 years old		6.54* (0.08)		9.32* (0.11)
66 years old and more		0.88 (0.003)		18.85 (0.07)
Education				
Basic education		–		–
Apprenticeship		−7.15 (−0.10)		9.00 (0.12)
General education school		−6.67 (−0.05)		12.06 (0.08)
High (professional) school		−5.69 (−0.07)		11.65 (0.14)
University		−7.39 (−0.06)		20.23** (0.16)
Family structure				
Couple living with children		–		–
Person living alone		20.48*** (0.25)		9.38* (0.10)
Couple living without children		−3.04 (−0.04)		0.38 (0.004)
Person without cohabiting partner living with child(ren)		2.61 (0.02)		0.68 (0.004)
Other family structures		1.32 (0.01)		−1.04 (−0.01)
Context of the residence				
Peripheral municipality		−0.24 (−0.003)		7.75 (0.10)
Peri-urban municipality		–		–
Suburban municipality		−0.80 (−0.01)		7.47 (0.09)
Small centre		6.98 (0.06)		14.92* (0.11)

Table 6.3 cont'd

	Mean distance Ego-Alters		Mean distance Alter-Alter	
Middle centre		−2.47 (−0.02)		17.29** (0.15)
Large centre		−2.35 (−0.02)		19.32** (0.13)
Constant	16.65***	9.90	25.32***	−5.71
R	0.15***	0.48***	0.16***	0.44***
R²	0.02***	0.23***	0.03***	0.19***
ΔR²		sig. < .01		sig < .01

Notes
*   p < .1.
**   p < .05.
***   p < .01.
a   Standardised regression coefficients are in brackets.

Table 6.4 (left column) shows that the activation of the mutual emotional support ties between the respondent and his or her network members is statistically associated with the respondent's commuting. The more the respondent commutes, the lower the proportion of the significant others sharing support with him or her is. The regression model shows that for each increase of 10 km in the respondent's commuting distance, the activation of the mutual support ties between the respondent and the significant network members decreases by 1.4 per cent.

Is this decrease in the proportion of activated significant others a direct consequence of the respondent's commuting or is it due to the fact that, when the respondent commutes, the network members are farther away from the respondent (indirect effect)? By adding different control variables to the regression model, including the mean distance between the respondent and the significant others, we observed that this distance does not influence the exchange of emotional support. It is not the fact that the respondent is far apart from his or her significant network members that cause a decrease in their activation; it is the fact that the respondent is a commuter. In other words, commuters are less likely to share emotional support with their significant others and this relation is not mediated through an effect of distance between the commuter and his or her network members. We also observed that young adults, women and people with a high school degree support each other proportionally more with their significant others. Conversely, people living alone, in big urban centres and, to a lower extent, those living in small centres declare that they support each other proportionally less with their significant network members. This last outcome shows that the morphology of the residence context has an impact on mutual support.

The study of the received and given support (Table 6.4, right-hand columns) enabled us to refine the analysis. Concerning the activation of the given support relations, women and young adults give support to a larger proportion of their network members. Conversely, people living alone, in *other* family structures (living with parents, flatmates, etc.) and in big centres support their significant others proportionally less. Neither commuting distance nor education influences the support given by respondents. On the other hand, people with high levels of education reported receiving emotional support from a higher proportion of significant others than respondents with low levels of education. Conversely, elderly people, commuters and people living in small centres receive support from a smaller proportion of their significant network members.

These results indicate that the respondent's commuting distance do not considerably influence the proportion of significant others to whom he or she gives emotional support. Commuting distance negatively influences the proportion of significant others who give him or her support. This deficit can be interpreted as an effect of the commuter's mobile living arrangement, since increasing time spent travelling may foster a weaker involvement in the activities with significant others (relatives, close friends).

The analysis of the activation of support ties between significant others (Table 6.5) shows that neither the respondent's commuting distance nor the significant others' mean commuting distance exert a significant influence.

On the other hand, the physical distance between the significant others' residences negatively influence the activation of the support ties. This result stems logically from the fact that significant others who were far apart from each other had a greater chance to know each other less or not at all and, therefore, support each other proportionally less. This analysis also indicates that the networks of male respondents between 51 and 65 years old present a stronger proportion of significant others supporting each other. The respondents' education level and the context of residence, however, do not have any effect on the emotional support exchanged between significant others.

Commuting and Social Integration: More Widespread Relations and Less Activated Support Ties

Based on our results, our initial hypothesis is confirmed: *commuters are more likely to develop a social network that is less anchored in contiguity, more spatially expanded and more discontinuous than non-commuters*. The longer a respondent's commuting distance is, the larger the distance between the respondent's place of residence and those of his or her significant others is. Further, the longer the commuting distance, the higher the mean distance between the residences of the significant others (see Figure 6.2). Therefore, commuting is a factor of transformation of social integration, of its local

Table 6.4 Regression analysis of the activation of the emotional support ties (in ‰) Ego-Alters on different variables related to the respondent (unstandardised regression coefficients)[a]

	Activation of mutual support ties Ego-Alters	Activation of support ties Ego	Activation of support ties given by Ego	Activation of support ties received by Ego
Commuting distance	-1.34** (-0.09)	-0.63 (-0.05)	-0.87 (-0.07)	-1.40** (-0.10)
Distance Ego-Alters	-1.43** (-0.09)			-1.26** (-0.09)
Sex (female)	-0.02 (-0.002)		-0.05 (-0.01)	0.19 (0.02)
	64.69** (0.09)		70.02** (0.11)	43.15 (0.07)
Age				
18–34 years old	85.84** (0.11)		75.64** (0.11)	54.02 (0.07)
35–50 years old	—		—	—
51–65 years old	41.92 (0.05)		-6.89 (-0.01)	44.26 (0.06)
66 years old and +	-103.84 (-0.04)		22.65 (0.01)	-194.48* (-0.08)
Education				
Basic education	—		—	—
Apprenticeship	77.50 (0.10)		52.30 (0.08)	59.58 (0.09)
General education school	-67.78 (-0.05)		-33.17 (-0.03)	-61.07 (-0.05)
High (professional) school	129.06** (0.16)		63.76 (0.09)	91.88* (0.13)
University	105.59 (0.09)		12.35 (0.01)	122.55** (0.12)
Family structure				
Couple living with child	—		—	—
Person living alone	-87.56** (-0.10)		-88.43** (-0.12)	-47.34 (-0.06)

Table 6.4 cont'd

	Activation of mutual support ties Ego-Alters	Activation of support ties given by Ego	Activation of support ties received by Ego
Couple living without children	−43.50 (−0.05)	−3.95 (−0.01)	−24.85 (−0.06)
Person without cohabiting partner living with child	0.70 (0.00)	8.84 (0.01)	−18.31 (−0.01)
Other family structures	−91.58 (−0.08)	−131.83***(−0.13)	−8.56 (−0.01)
Context of the residence			
Peripheral municipality	−36.90 (−0.04)	57.14 (0.08)	−16.06 (−0.02)
Peri-urban municipality	–	–	–
Suburban municipality	−16.65 (−0.02)	33.16 (0.05)	14.67 (0.02)
Small centre	−119.89* (−0.10)	−16.07 (−0.02)	−147.25*** (−0.13)
Middle centre	16.15 (0.01)	49.44 (0.05)	12.40 (0.01)
Large centre	−135.85**(−0.10)	−107.29* (−0.10)	−85.57 (−0.07)
Constant	681.99***	783.11***	735.40***
R	0.26***	0.28***	0.26**
R²	0.07***	0.08***	0.07**
ΔR² a	sig. < .05	sig. < .01	sig. < .05

	Activation of mutual support ties Ego-Alters	Activation of support ties given by Ego	Activation of support ties received by Ego
Constant	744.6***	853.8***	802.9***
R	0.09**	0.05	0.09**
R²	0.01**	0.002	0.01**

Notes: * p < .1; ** p < .05;*** p < .01.
a Standardised regression coefficients are in brackets.

Table 6.5 Regression analyses of the activation of the emotional support ties (in ‰) Alter-Alter on different variables related to the respondent (unstandardised regression coefficients)[a]

	Activation of support ties Alter-Alter	
Commuting distance	−1.11 (−0.08)	−0.66 (−0.05)
Mean commuting distance of the Alters		0.59 (0.03)
Distance Alter-Alter		−1.09** (−0.12)
Sex (female)		−97.60** (−0.14)
Age		
18–34 years old		−7.29 (−0.01)
35–50 years old		–
51–65 years old		110.01** (0.13)
66 years old and more		−50.47 (−0.02)
Education		
Basic education		–
Apprenticeship		5.41 (0.01)
General education school		−17.71 (−0.01)
High (professional) school		21.61 (0.03)
University		58.24 (0.05)
Family structure		
Couple living with children		–
Person living alone		22.66 (0.03)
Couple living without children		−49.72 (−0.06)
Person without cohabiting partner living with children		27.91 (0.02)
Other family structures		−54.95 (−0.05)
Context of the residence		
Peripheral municipality		−49.25 (−0.06)
Peri-urban municipality		–
Suburban municipality		−75.33 (−0.10)
Small centre		−9.48 (−0.01)
Middle centre		−28.09 (−0.03)
Large centre		83.17 (0.06)
Constant	401.68***	475.14***
R	0.08	0.27
R^2	0.01	0.07
ΔR^2		n.s.

Table 6.5 cont'd

Notes
* $p < .1$.
** $p < .05$.
*** $p < .01$.
a   Standardised regression coefficients are in brackets.

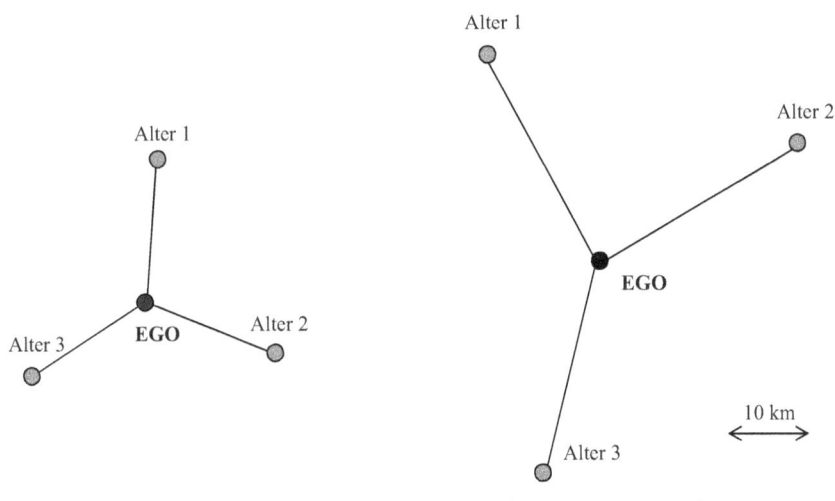

a) Ego is weakly commuting   b) Ego is strongly commuting

Figure 6.2   Illustration (to scale) of the network spatial expansion according to a weak (2 km) or strong (50 km) commuting distance of the respondent

Note: Distances were drawn to scale, based on the predictions of the simple regression model.

embedding and of its recomposition on a larger scale. The commuter becomes the centre of a spatially widened network whose members are more distant from each other than traditional social networks.

Thus, commuting practices must not be strictly understood as a way to maintain a locally embedded and densely connected social network, but also as a mobile living arrangement fostering a spatially expanded social anchoring. These spatial recompositions of social integration have a series of relational consequences. Previous research has shown that the frequency of interactions is very sensitive to geographical distance, which hinders contacts and exchanges (see, for instance, Coenen-Huther et al. 1994; Bonvalet and Maison 1999; Axhausen and Frei 2007). Therefore, some important ties tend to become virtual, or at least, to become more potential than active.[6] It is what we observed if we consider the activation of the mutual emotional support ties between an individual and his or her significant others that weakened when the individual's commuting distance increased. Thus, spatial distance has relational repercussions because it integrates commuters in interpersonal relationship networks in which the proportion of non-activated significant persons is higher. In particular, this is the proportion of network members supporting the commuter that decreases when the distance from home to work increases, whereas the proportion of network members receiving some support from the commuter remains stable. In accordance with our hypothesis, we did not measure any significant differences in the number of support ties according to commuting. The commuter, in particular the long distance commuter, tends to quote more significant network members, even if they are proportionally less activated in their support with him or her. Commuting is indeed associated with a structural recomposition and not a weakening of interpersonal relationships. Relational anchorings associated with commuting are as important, if not more important, than others, but potentially less supportive.

It is, therefore, vital to investigate the extent to which emotional relations develop differently within a more spatially expanded and more discontinuous network. Some forms of emotional support, those forged in habit and daily time, can be more difficult to share with commuters, given that they are relationally anchored in different places. Time spent travelling may also be a hindrance to more involvement in social life. These elements could partially explain why commuters claim to receive less support than they provide. On the other hand, commuting may foster the constitution of close relations, with colleagues for instance, who are disconnected from the rest of the solidarity network because of physical distance between home and workplace. Other forms of support within the primary network (e.g., confidence relationships, professional relationships) could then develop. At last, we can also assume that commuters may try to develop, through the physical distance from their social relations, different solidarity dynamics, based for instance on a lower level of responsibility towards them (e.g., children, relatives).

By the mediating effect of the physical distance, commuting is moreover associated with a higher proportion of significant others not supporting each other. The significant others support each other less, probably because they do not know each other well and have fewer opportunities to be together. Owing to a more widespread social anchoring, the commuter cannot fully exert this logic of transitivity, which is at the core of the social networks construction (Widmer 1999). This results in a situation of *bridging social capital*, in which

---

6 Because the frequency of interactions between network members was not collected in the MOSAiCH survey, this proposition remains on the order of a hypothesis.

the commuter becomes the compulsory intermediary between the members of his or her network (Burt 1992, 2002). If the respondents themselves present less activated support ties with their significant others in situations of commuting, this suggests the presence of weaker ties, confirming again the constitution of a bridging social capital. This has important potential consequences. Taking advantage of greater autonomy because their significant others are sparsely connected to each other, commuters can benefit from intersecting social circles (Simmel 1999). Through this relatively new social integration that is particular to modernity, they can develop an original identity, a sort of synthesis of various influences, which are physically distant and relatively disconnected from them. Because of a position of intermediary between disconnected individuals, commuters take advantage of various and non-redundant materials and resources and can play the role of mediator, controlling exchanges between their significant others (Burt 1992). Additionally, they are not constrained by closed networks (Coleman 1988) characterised by a strong normative pressure (i.e., everybody knows everybody in the network and all members react collectively to rumoured or real deviances). However, on the other hand, commuters do not benefit from the collective activation of a set of interconnected persons, where trust and mutual aid are reinforced by the collective constraint.

The results discussed above highlight the net effect of commuting from the effects of other variables, such as education, sex, age, family structure, residential mobility and residence context. In other words, if commuting exerts a negative effect on the activation of the support ties, it is not because commuters are mainly men, well-educated, or inhabitants of urban centres. Other variables create their own important effects. First, the distance between the residence at the age of 14 and the current residence produce very similar effects to commuting on the network spatiality. If this implicitly suggests that social ties are progressively built since early childhood, it also shows that migrations, inside or outside[7] the country, have opened the traditional modes of social integration and recomposed them in a broader space. This influence of the residential trajectories was notably brought to light by the studies on the spatiality of family configurations (Bonvalet and Maison 1999; Bonvalet and Lelièvre 2005). Because the two mobility dimensions, residential and professional, are going to increase and reinforce each other, we see what their joint effect on interpersonal relationships could be. The two mobility dimensions could result in networks that, without being smaller, will be less and less dense and more widespread, showing a *bridging* logic.

Among other important results, residence context is significant. Our analyses show in this respect that commuting influences social relationships differently according to the context of a respondent's residence. For equivalent commuting, the inhabitants of urban centres have social networks, which are,

---

7   External migration was not measured in this study.

at the same time, more widespread and characterised by a lower activation of mutual support ties with their significant others than people living in suburban, peri-urban, or peripheral municipalities. The urban morphology, i.e., the *visible* dimension of the city continues to be a social marker. Contrary to a now dominant discourse on the urban question, the city has not been totally dissolved in even broader conurbations with even blurrier borders. This result particularly shows that the relational anchorings of city centre inhabitants differ from those of people living in the city outskirts. This observation is not reduced to the different composition of the population, in terms of education or family structure. This social integration, characterised by less dense and more widespread social networks, can also be explained by the stronger presence of foreigners in urban centre contexts.[8]

Added to this, the educational level and the residence context present a very interesting effect on network spatiality. These factors only influence the distance between the significant others, whereas the distance between the respondent and the network members hardly varies. Highly educated people living in urban centres benefit from a network where they are more centrally located according to the spatial position of their significant others. Conversely, less educated people living outside the urban centres are less centred. Additionally, this result supports the thesis that less educated people living in the outskirts are more likely to find themselves farther from their significant network members than they are from each other (see Figure 6.3).

The more reticular living spaces that were highlighted in this chapter imply that the potential to be mobile, i.e., *motility*, becomes an essential element from which social networks are built and maintained. In a general way, motility may be defined as the manner in which an individual appropriates the field of possibilities relative to movement and uses them (Kaufmann 2002; Kaufmann et al. 2004). In our study, it may be more precisely understood as a potential or real capability to maintain significant support ties in spite of physical distance; to keep in touch by means of possibilities offered by the transportation and communication systems; and the ability to build new significant relationships in various places. A strong mobility capital allows individuals to maintain or widen their social capital. The existing literature on the domain shows that these capabilities are not egalitarian over the social structure. For the most disadvantaged population categories,[9] several factors

---

8 The proportion of respondents who lived abroad at the age of 14 and live currently in a urban centre reaches 23 per cent against 16 per cent for those living currently in a suburban, peri-urban or peripheral municipalities. Given that their municipality of residence at the age of 14 is outside the Swiss territory, the distance of earlier residential mobility cannot have been measured and was then defined as missing values. These individuals have thus not been identified as having a strong residential mobility.

9 In particular, isolated women with children, migrants, less educated and disabled people.

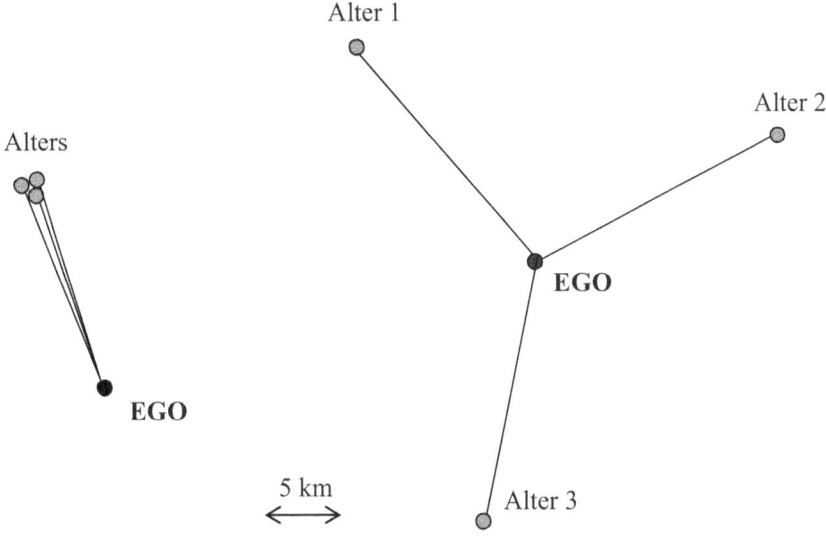

Figure 6.3   Illustration (to scale) of the network spatial expansion according to the education level and residential context of the respondent

Note: Distances were drawn to scale, based on the predictions of the multiple regression model. The other variables were fixed identically in both situations (commuting distance: 10 km; distance between the current residence and the one at the age of 14: 20 km; male; 35–50 years old; family structure: couple living with children).

may intensify their difficulties to maintain significant social relations in daily life. Having no car (Dupuy 1999; SEU 2002 report cited in Urry 2007, 13; Gray et al. 2006); living in a residence apart from transportation facilities and meeting places (shops, bars) (Church et al. 2000; Cass et al. 2005; Kenyon 2006); or weak resources, in organisational or temporal terms, to travel in order to see friends and relatives (Kaufmann et al. 2005; Le Breton 2005; Urry 2007) may explain such difficulties.

Such social inequalities regarding motility can explain differences in the proportion of supporting significant others according to education and commuting. Though highly educated people and commuters present a broader relational space, the former are more likely to have a higher proportion of significant others supporting them than less educated people. Conversely, commuters are more likely to have a lower proportion of supporters compared with non-commuters. Highly educated people have a stronger ability to create and maintain support ties with significant people who are not physically close than less educated people do. The spatial fragmentation between home and workplace, as well as time spent in transportation, could explain the reversed result by commuters. In more general terms, the broader spatiality of social anchorings has consequences on the way to tackle the issue of socio-spatial inequalities. It is notably hazardous to measure social segregation in an agglomeration from the only residential location. Because the residence municipality is not necessarily the centre of social life any more, a segregation measure must take into account the social anchorings realised in a more broadly space.

Conclusion

This chapter referred to some of the dimensions associated with the new spatiality of social integration. Several issues remain open at this stage and further explorations should clarify them. Questions arise about the impact of the network composition on its spatiality. If, for example, people living alone have networks which are more spatially fragmented and relationally less dense, it may be because they have no cohabiting partner. Conversely, perhaps some people have more locally anchored networks because they live with children who are old enough to be mentioned within their network. As suggested by the typology developed by Wellman et al. (1988), commuters might correspond to these modern individuals, having a physically widespread *liberated community* of friends and colleagues, whereas relatives might remain embedded in a more local community. It would also be worth exploring the influence of family recompositions or municipalities' accessibility on the spatial expansion of social relationships. Finally, by focusing on support ties, we weighted the emotional dimension of social ties, favouring strong ties compared with weak ties. By concentrating on a relationship form more characteristic of weak ties (relationship as information channel, influence relationships, more superficial discussions), bridging social capital might be shaped more markedly in situation of commuting.

The links between geographic mobility and social capital highlighted in this chapter should not be understood merely as an univocal effect of the first factor on the second one. Dynamics between spatial dimension and relational dimension are certainly more interactive; both dimensions may reinforce each other over the life course. If high mobility fosters a more widespread social network, this latter may lead to new forms of spatial mobility, given the less localised relational anchoring. A process of cumulative effects (Dannefer 1988) may then occur: small differences in the social network and the mobility experiences of one individual, when they combine, can produce very different life trajectories. Therefore, relocating in the first part of life may lead to a spatially and relationally more discontinuous social network, which in turn may foster a stronger willingness for new experiences of spatial mobility.

These analysis dimensions must still be scrupulously studied, but our overall finding is nevertheless solid. By favouring spatial mobility, modernity creates new means to be relationally anchored. The example of commuting that we developed in this chapter shows in particular that integration modes become relationally less dense and that space, within which social networks are established, can be, at the same time, very distant from the residence and fragmented.

References

Axhausen, K.W. and Frei, A. (2007), 'Contacts in a Shrunken World', *Working papers, Verkehrs- und Raumplanung, 440, IVT, ETH Zürich*. Zürich.
Boltanski, L. and Chiapello, E. (1999), *Le nouvel esprit de capitalisme* (Paris: NRF, Gallimard).
Bonvalet, C. and Maison, D. (1999), 'Familles et entourage: le jeu des proximités', in Bonvalet C. Gotman, A. and Grafmeyer, Y. (eds), *La famille et ses proches, l' aménagement. des territoires* (Paris: PUF).
Bonvalet, C. and Lelièvre, E. (2005), 'Les lieux de la famille', *Espaces et sociétés* 1-2, 99-122.
Bourdieu, P. (1980), 'Le capital social: notes provisoires', *Actes de la Recherche en Sciences Sociales* 31:1, 2–3.
Breton, E. le (2005), *Bouger pour s'en sortir. Mobilité quotidienne et intégration sociale* (Paris: Armand Colin).
Burt, R. (1992), *Structural Holes: The Social Structure of Competition* (Cambridge, MA: Harvard University Press).
Burt, R. (2002), 'The Social Capital of Structural Holes', in Guilléen, M.F., Collins, R., England, P., and Meyer, M. (eds).
Cass, N., Shove, E. and Urry, J. (2005), 'Social Exclusion, Mobility and Access', *Sociological Review* 53:3, 539–55.
Chamboredon, J.-C. and Lemaire, M. (1970), 'Proximité spatiale et distance sociale. Les grands ensembles et leur peuplement', *Revue française de sociologie* 11:1, 3–33.
Church, A., Frost, M. and Sullivan, K. (2000), 'Transport and Social Exclusion in London', *Transport Policy* 7:3, 195–205.
Coenen-Huther, J., Kellerhals, J. and von Allmen, M. (1994), *Les réseaux de solidarité dans la famille* (Lausanne: Réalités sociales).
Coleman, J. (1988), 'Social Capital and the Creation of Human Capital', *American Journal of Sociology* 94:Suppl., 95–121.
Dannefer, D. (1988), 'Differential Aging and the Stratified Life Course: Conceptual and Methodological Iissues', in Maddox, G. and Lawton, P.M. (eds), *Annual review of Gerontology* (New York: Springer).
Dupuy, G. (1999), *La dépendance automobile* (Paris: Anthropos-Economica).
Frei, A. and Axhausen, K.W. (2007), 'Size and Structure of Social Network Geographies', *Arbeitsberichte Verkehrs- und Raumplanung, 439, IVT, ETH Zürich*.

Frick, R. (2004), *'Recensement fédéral de la population. La pendularité en Suisse'* (Neuchâtel: Office Fédéral de la Statistique).

Granovetter, M.S. (1982), 'The Strength of Weak Ties. A Network Theory Revisited', in Mardsen P. V. and Lin, N. (eds), *Social Structure and Network Analysis* (London: Sage).

Granovetter, M.S. (2000), *Le marché autrement: les réseaux dans l'économie* (Paris: Desclée de Brouwer).

Gray, D., Shaw, J. and Farrington, J. (2006), 'Community Transport, Social Capital and Social Exclusion in Rural Areas', *Area* 38:1, 89–98.

Guilléen, M.F., Collins, R., England, P. and Meyer, M. (eds) (2005), *The New Economic Sociology: Developments in an Emerging Field* (New York: Russel Sage Foundation).

Hollstein, B. and Straus, F. (eds) (2006), *Qualitative Netzwerkanalysen. Konzepte, Methoden, Anwendungen* (Opladen: VS Verlag).

Jemelin, C., Kaufmann, V., Barbey, J. and Pflieger, G. (2007), 'Inégalités sociales d'accès: quels impacts des politiques locales de transport?', EspacesTemps.net, <http://www.espacestemps.net/document2263.html>, accessed 25 May 2008.

Kaufmann, V. (2002), *Re-thinking Mobility* (Aldershot: Ashgate).

Kaufmann, V., Bergman, M.M. and Joye, D. (2004), 'Motility: Mobility as Capital', *International Journal of Urban and Regional Research* 28:4, 745–65.

Kaufmann, V., Montulet, B. and Le Breton, E. (2005), 'Mobilité et mobilités familiales', *Netcom – Networks and Communication Studies* 19:3, 137–39.

Kennedy, P. (2004), 'Making Global Society: Friendship in Networks among Transnational Professionals in the Building Design Industry', *Global Networks* 4:2, 157–79.

Kenyon, S. (2006), 'Reshaping Patterns of Mobility and Exclusion? The Impact of Virtual Mobility upon Accessibility, Mobility and Social Exclusion', in Sheller M. and Urry, J. (eds), *Mobile Technologies of the City* (London: Routledge).

Kesselring, S. (2005), 'New Mobilities Management. Mobility Pioneers between First and Second Modernity', *Zeitschrift für Familienforschung* 17:2, 129–43.

Kesselring, S. (2006a), 'Topographien mobiler Möglichkeitsräume. Zur soziomateriellen Netzwerkanalyse von Mobilitätspionieren', in Hollstein, B. and Straus, F. (eds), *Qualitative Netzwerkanalysen. Konzepte, Methoden, Anwendungen* (Opladen: VS Verlag).

Kesselring, S. (2006b), 'Skating over Thin Ice. Pioneers of the Mobile Risk Society', paper presented at 'Reprendre Formes. Formes urbaines, pouvoirs et expériences. Séminaire international de réflexion en présence de Manuel Castells', June 2006, Lausanne.

Maddox, G. and Lawton, P.M. (eds) (1988), *Annual review of Gerontology* (New York: Springer).

Mardsen P. V. and Lin, N. (eds) (1982), *Social Structure and Network Analysis* (London: Sage).

Offner, J.-M. and Pumain, D. (eds) (1996), *Réseaux et territoires. Significations croisées* (Paris: Éditions de l'Aube).

Ohnmacht, T., Frei, A. and Axhausen, K.W. (2008), 'Mobilitätsbiografie und Netzwerkgeografie: Wessen soziales Netzwerk ist räumlich dispers?', *Swiss Journal for Sociology* 31:1, 131–64.

Portes, A. (1998), 'Social Capital: Its Origins and Applications in Modern Sociology', *Annual Review of Sociology* 24:1, 1–24.

Putnam, R. (2000), *Bowling Alone* (New York: Simon and Schuster).

Scheller, M. and Urry, J. (eds) (2006), *Mobile Technologies of the City* (London: Routledge).

Simmel, G. (1999), *Sociologie, Étude sur les formes de la socialisation* (Paris: PUF).

Urry, J. (2000), *Sociology beyond Societies. Mobilities for the 21st Century* (London: Routledge).

Urry, J. (2007), 'Des inégalités sociales au capital en réseau', *Swiss Journal of Sociology* 33:1, 9–26.

Wellman, B., Carrington, P. and Hall, A. (1988), *Networks as Personal Communities in Social Structures: A Network Approach* (New York: Cambridge University Press).

Widmer E.D. (1999), 'Family Contexts as Cognitive Networks: A Structural Approach of Family Relationships. Personal Relationships', *Special Issue on Methodological and Data Analytic Advances in the Study of Interpersonal Relationships* 6:4, 487–503.

Widmer E.D. (2006), 'Who are my Family Members? Bridging and Binding Social Capital in Family Configurations' *Journal of Personal and Social Relationships* 23:6, 979–98.

Wilson, W.J. (1987), *The Truly Disadvantaged* (Chicago: University of Chicago Press).

Wittel, A. (2001), 'Toward a Network Sociality', *Theory, Culture and Society* 18:6, 51–76.

This page has been left blank intentionally

Chapter 7
Here, There, and In-Between: On the Interplay of Multilocal Living, Space, and Inequality

Nicola Hilti

This chapter deals with the broad and intangible phenomenon of multilocal living, which becomes more and more important throughout late-modern Western societies. Multilocal living as it is understood in this context is a way of organising everyday life in and between different homes. Multilocalists have more than one place of residence; at more or less regular intervals they live (and work) in these different places.[1] Against the backdrop of a series of actual societal processes, multilocal living has become an increasingly realised option.

Multilocal living provides the structuring framework (of opportunities and constraints) for everyday life and is also an everyday practice. Therefore it influences various private spheres: housing, work, leisure, family and other social relations. Multilocal living touches fundamental social aspects of politics, economics, infrastructure, housing markets etc.

Prior to tackling the core issues of this chapter – i.e. the links between multilocal living, space and inequality – I shall briefly introduce the state of research and put forward some notes on the conceptual framework of my empirical research. I shall follow this with a series of examples illustrating the range and diversity of multilocal lifestyles.

Mobility (with its spatial and cultural dimensions) – a major key word of modernity – is the very precondition of multilocal living. By their mobility within and between different places, multilocalists influence and configure space in a specific way – and are themselves influenced and 'configured' by it. To show this link between multilocal living and space, I shall describe one main premise in discussions of multilocal living, i.e. the close interplay between mobility and settledness. They can be considered as two sides of the same coin, with multilocal living situated right between the two spheres. In other words: multilocal living can be described as a semantic field of tension between the

---

1 In academic literature you sometimes find terms like *multiresidentiality* (Kaufmann 2002), *multiple residency* (Luka 2006) or *habiter pluri-topique* and *habiter multi-topique* (both: multi-topical dwelling) respectively (Stock 2006).

desire and need to be mobile and the wish to be settled. Against this backdrop *the in-between, the interstices*, are identified as crucial for studying the field of multilocal living (besides other spatial and temporal aspects, which will not be the focus of this chapter). Interstices have a specific meaning in the context of a multilocal organisation of everyday life, and they are closely linked to the topic of *inequality*. I shall exemplify the complex interfaces between interstices and inequality chiefly by one of the most relevant means of transport (besides the car) for multilocalists in Central Europe: the railway. Furthermore, social inequalities (can) lead to different options and constraints for multilocalists. As today's multilocal living as such is not a matter of social class or lifestyle anymore and thus has entered societal mainstream, it is vital to study those links to questions of inequality. Multilocality in its various forms runs through all social categories, as Bonnin and de Villanova (1999, 15) state. Hence, there are different degrees of freedom in deciding upon and coping with multilocal living. The social inequality related to the lifestyles of multilocalists is also represented by the amount of (scientific) attention which different forms of multilocal arrangements get.

Current State of Research

Even though a number of scientists from different disciplines have looked at mobile people, the phenomenon of living in more than one place is treated either a minor topic (e.g. Kaufmann 2002) or limited to single forms of mobile (and implicitly multilocal) lifestyles or single aspects of multilocal living such as the motives of job-related mobility and its effects on family life (e.g. Schneider et al. 2002a, 2002b; Collmer and Kümmel 2005; Zvonkovic et al. 2005). Most studies focus on the middle or upper classes and on highly educated professional groups, e.g. professionals representing financial, media and creative industries (e.g. Englisch 2001; Pelizäus-Hoffmeister 2001; Bonss et al. 2004; Kreutzer and Roth 2006). Extensive research on multilocalists such as truck drivers, permanent campers or people with allotment gardens is rather difficult to find. Likewise only a few authors focus on the individual's perspective on mobility desires and constraints (e.g. Albrow 1997; Bonss 1999; Wallace et al. 1999; Vonderau 2003). Although mobility (and motility[2] respectively) is identified as a relevant factor of social inequality (Kaufmann

---

2 The term *motility* stems originally from biology and medicine where it is used to describe the capacity of an organism to move. Motility in social sciences and transportation research means 'the capacity of entities (e.g. goods, information or persons) to be mobile in social and geographic space, or as the way in which entities access and appropriate the capacity for socio-spatial mobility according to their circumstances' (Kaufmann et al. 2004, 750).

et al. 2004; Kaufmann et al. 2007) research on different forms of mobility and their potential for deepening or minimising inequalities is still in its infancy.

Second homes are mainly dealt with in the context of tourism and its economic and social effects (e.g. Gallent and Tewdr-Jones 2000; Hall and Müller 2004), and of leisure mobility and traffic behaviour (e.g. Fuhrer and Kaiser 1994). One can also find regional registrations and typologies of second homes (e.g. Xiao Di et al. 2001). Research on tourist practices still concentrates on the destination and rarely on the interaction between the places of home and vacation and its effects (e.g. Urry 1990; Löfgren 1999; Opaschowski 1999; Rolshoven 2003a).

A similar situation can be found in migration research: in general, problem-oriented questions of destination countries dominate. Only recently have sociologists and cultural anthropologists in particular started to focus on the interaction between places of origin and destination (e.g. Schiffauer 1991; Albrow 1997; Beck, 1998).

Other research tackles questions of belonging to (home) places and the field of tension between freedom and constraint framed by mobility (e.g. Kaiser 1993; Tully and Baier 2006), or the decision-making processes with regard to different forms of mobility and their consequences (e.g. Kalter 1994; Kecskes 1994; Jürges 1998; Schneider et al. 2002a, 2002b). One can also find basic critical arguments concerning mobility constraints in a globalised and capitalist world (e.g. Beck, 1986, 2007; Meier 1999; Sennett 2000), often implicitly lamenting an increasingly helpless and lost individual. The perspective on problems still dominates scientific discourse. Only a few studies also identify mobility as a (possible) factor of productive cultural change (e.g. Appadurai 1991; Hannerz 1998; Werbner 1999; Yeoh 2003).

Research on the multiplication of individual social and spatial relations usually does not reflect the complexity of multilocal living in everyday life (Rolshoven 2002a, 2002b). Despite the fact that close interrelations between mobility and settledness are commonly identified as highly relevant (e.g. Guzzoni 1999; Breckner 2002; Selle 2002; Merkel 2003; Rolshoven 2004; Schröer 2006), they are rarely considered in empirical research. Studies focusing on the physical and/or virtual space between places, emphasise the dynamic character of the relationship between structural conditions and individual actions (e.g. Augé 1994; Marcus 1995; Wicker 1996; Beck, 1997; Römhild 2003; Welz 2003; Widmer 2003; Axhausen 2007) and the necessity of a historical perspective (e.g. Löfgren 1995).

Multilocal Living as a Form of Multilocality

In anthropological research the concept of multilocality has already been in use for several decades (e.g. by Susan Watts in the 1970s and earlier by her colleague R. Mansell Prothero in the context of studying geographic population circulation in West Africa[3]). But in research on social change and its implications on Western societies it is a rather new concept. The term as I understand it, is (and ought to stay) generally dynamic and open. It seems neither possible nor desirable to find or to freeze a concrete and widely valid definition of multilocality. Therefore it makes sense to define and limit it according to each specific research context. Nonetheless, all possible interpretations seem to have one feature in common: by their multilocality people connect different meaningful places and build up individual meaningful arrangements. For example, the idea of multilocality 'as *vita activa* in several places' (Rolshoven 2006, 181 [translation by the author]) is a more philosophical interpretation, which largely foregoes formal definitional restrictions. The term is wide open and describes multilocality as 'the fullness of everyday activity ... distributed across several locations that are visited at longer or shorter intervals and used in a looser or stricter division of functions' (ibid. [translation by the author]), a definition, that implies 'multilocality as a way of life' (Rolshoven 2007) which affects almost any member of our society. For empirical research the problem of lack of delimitability rears its ugly head here.

Therefore, in my work I use *multilocal living* as a specified form of the overall concept of *multilocality*. Multilocal living implies the existence and use of more than one place of residence. This usage can follow different rhythms and motives. These places of residence can take on very different forms and qualities. Living as having a home assumes staying overnight (see examples below). Multilocal living is defined as part and parcel of the way everyday life is organised (as opposed to the yearly around-the-world-trip for instance). As my research is still in progress, this definition is deliberately vague. It complies with the open approach of Grounded Theory and helps to sharpen the senses for unexpected findings. For the final in-depth analysis in my research I hope to create a typology for differentiating various forms of multilocal living.

The Broad Variety of Multilocal Living

In order to make the phenomenon more tangible, I shall describe below some further examples, which illustrate the sheer range of forms and origins of multilocal lifestyles. In addition, it shows that multilocal living occurs in numerous professional groups and many different types. It is not merely about job-related or leisure-related multilocal living. If one thinks of a multilocal way of life, three or four typical forms will probably come to mind at first, several others are likely to appear surprising: *the couple temporarily living apart*, she works in Zurich, he has a job in Berne, and therefore he rents another apartment there – in addition to the shared domicile in Zurich. *The*

---

3 This fact has been investigated by Justin Winkler in the frame of his research on the term of multilocality.

student who rents a room in Lausanne, where she studies; on weekends she stays at her parents' house in the Rhine Valley. *The Ticino enthusiast*, she lives in Basle and works in Zurich, but spends every weekend at the house by the lake south of Switzerland. Eventually one would also think of *the retired couple* spending the winter months in their house on the Spanish Costa Blanca. One could also think of other forms, e.g. *the truck driver* who regularly sleeps in his cabin, *the musician* staying in ever-changing hotel rooms, *the child of divorced parents* who has a room at each of his or her parents' homes, *the grandmother* looking after her grandchild and therefore spending two nights a week at her daughter's house, *the nurse* who has a room at the hospital's boarding house, *the pilot* who regularly spends nights away from home (even *rail conductors* have sleeping quarters at the different railway stations), or *the seasonal labourer* whose rhythm between here and there is determined by various growing seasons, *the Member of the Parliament* who lives in a hotel or an apartment in Berne during parliamentary sessions, *the permanent camper* spending every weekend at the camping ground or even living there for the summer, and finally even *the mountain farmer* with his three-step- farming system[4] could be seen as multilocal. Many more examples could be quoted.

Some of these examples show that multilocal living as such is not new. In former eras too there was a range of forms of living in and between different homes and/or places (the other places of living were not necessarily felt as 'home'): e.g. (semi-)nomads, sailors, commercial travellers, educational and cultural travels of the upper class, summer stays in the countryside etc. What we observe today, however, are remarkable quantitative and qualitative shifts concerning the phenomenon of multilocal living within recent decades. We can assume that pre-modern forms of multilocal living (e.g. farmers managing and thus living in different places or children seasonally sent to farming families) decrease, while late-modern variants increase. Other long-established forms are still highly relevant (e.g. migrant workers).

Thus multilocal living as such is not a product of modernity or late-modernity, but has changed remarkably in the light of a number of, historically speaking, relatively recent social, political and economical developments: the flexibilisation of work, processes like individualisation or female emancipation, the diffusion of novel information and transport technologies, the emergence of mass tourism, to name but a few. The specific relationship between pre- and late-modern forms of multilocal living is a highly complex issue and goes beyond the thematic scope of this chapter.

---

4 This farming system is practised in western Austria for instance. The farmer's family and their cattle move through three altitudinal levels in a seasonal cycle as the pastures come to maturity.

Methodical Approach

The key interest of my work centres around the questions: how do multilocalists cope with their specific situation? What are the underlying intentions and meaning structures? What does multilocal living mean for feelings of belonging to places and groups? My main goal is to give an initial overview of a phenomenon so far largely ignored in research. Therefore I am not only interested in one specific form but rather in the whole spectrum of multilocal living. The research is situated in the national context of Switzerland.

Empirical evidence is being collected by qualitative research. In order to grasp the enormous variety and high complexity of the phenomenon, I am gathering multilocalists' perspectives via problem- or topic-centred interviews (including biographical and narrative elements) (Schlehe 2003), complemented by short questionnaires and some field observations, e.g. during a stay on a camping ground in Basle. Additionally, a variety of discursive sources will be discussed. Interviews with 25 multilocalists had been conducted up to the time of writing. Most of them have at least one of their homes in Switzerland, which does not mean that multilocal living is specifically Swiss. The phenomenon as such seems to be found in all Western societies and possibly beyond. Of course there are differences depending on the relevant national political, economical, social and cultural background, e.g. the culture of using second homes in Scandinavia is different from that in France.

The empirical research is Grounded Theory-oriented (Glaser and Strauss 1968), an approach which offers specific procedures for collecting and analysing empirical data to build up new theories. The different research phases overlap, so that new cases are chosen following the analysis of the first interviews. At first I looked for interview partners rather unsystematically and in different networks independent from each other. Later on, my efforts became more focused: I contacted specific gatekeepers, e.g. to get on a camping ground, or I wrote to people who had placed postings on internet portals such as 'homesick Appenzeller looking for a second home'. I am aiming for structural variance, which reflects the heterogeneity of the field (rather than its statistic representativeness). The data analysis is mainly following suggestions which combine reconstructive and descriptive steps (e.g. Froschauer and Lueger 2003).

Mobility, Settledness and Interstices

Two Sides of the Same Coin

The major prerequisite for any idea of space is movement. Movement constitutes spaces (and borders) (Löfgren 1995, 359–60). Daily actions and daily paths take place in a certain space and therefore generate space as such

(Rolshoven 2003b). Movement in turn leads to the concept of mobility. And, mobility is the very precondition of multilocal living. The multilocalist is not only a resident but also a mobile person. He or she at times lives at one certain place and at other times somewhere else – not to forget that he or she is (often) on the road, in transit or in-between (these places). Mobility is the counterpart of settling down and being in situ. Ina Merkel postulates that mobility and settledness do not exist solely for themselves, they are always corresponding to each other and therefore never exclusive (Merkel 2003).

Thus, multilocal living is situated right between mobility and settledness.[5] And it is also the very link between these two spheres. In this light multilocal living can be understood as an actively chosen and lived compromise between different needs that cannot be satisfied in one place. There is more than one place, which fulfils more or less differentiated functions. This is why multilocal living is not about being rootless, but about having more than one place of residence or more than one place that people call 'home'. Empirical research provides strong support for this observation: multilocalists maintain different forms of anchors guaranteeing (a certain) constancy (in the sense of caring about roots). For example, the father and his house are a constant in the life of the young multilocalist with a distinctly mobile lifestyle. The father and the house constitute a consistent and always open base station should any problems occur. This reliable fall-back option is very important to the multilocalist, even if she would never really make use of it. This place is perceived as something static and constant, a place for recharging the batteries, a source of power for 'the other life' (Löfgren 1995, 356), while the mobile lifestyle between the other places is seen as a dynamic counter concept to a lifestyle described in terms of a standstill, treadmill sort of everyday life for couch potatoes. Once again we see that having a residence and mobility are closely linked. Orvar Löfgren argues that even though we claim to be mobile and flexible, we still long for a place where we belong (Löfgren 1995, 356). Thus multilocal living can also be a strategy to maintain or realise such a place of 'home' (as well as to maintain a social status, as argued below).

As we noticed, the desire for constancy can be expressed by the significance of having or creating anchors as a counterbalance. But the wish for constancy or invariability can even be the prime reason for a multilocal way of life. Löfgren emphasises – critically against today's popular ideas of de-territorialisation – that mobility does not necessarily lead to loss of place-relatedness. Moreover,

---

5   The close connection between mobility and settledness is somehow true for most people, who reside somewhere, but is manifest in multilocal living in a very specific way. Depending on the individual circumstances possible exceptions would be: closed ward psychiatry patients, prisoners and the like. Above all, even homelessness should be considered in the context of multilocal living. The lack of having a home, a place where you belong, is conceived as a basic deficit in our society – be it on a spatially manifest or on an imaginary level. The term *homeless* already implicates this.

movement can even be the means of establishing constancy (Löfgren 1995, 352). One multilocal couple I met shares two apartments in Berne, which are located within a 15-minute bus ride of each other. From Monday to Friday they live in the Old Town, the weekends are spent in the other district. Two striking reasons for this arrangement were the man's physical handicap as well as his strong emotional rootedness and social integration in his residential area. Paradoxically one could say that this couple's multilocal living is based on a severe physical and spatial immobility.

The geographical distance is not a significant criterion for immersion in a 'counter world', as indicated by Michael C. Hall and Dieter Müller, who write that 'most second-home owners live close to their property' (Hall and Müller 2004, 8). This fact can also be seen very clearly in the phenomenon of permanent camping: permanent campers feel a big difference in relaxing and switching off regarding the start and end of their workdays on the camp site as compared to living in their apartments. This begs another interesting question: do multilocalists 'multiply' their life or do they seek to live 'different lives' in different places? We can assume that this question refers to two different types of multilocalists, with several more potential types of strategies.[6]

The strong interrelatedness of mobility and settledness is the reason why we cannot study multilocal living by simply focusing on the different places, but need also look at *the in-between* of places, *the interstices* – in German, *das Dazwischen* (Rolshoven 2004). Of course the conditions and activities *within* the different contexts of the living places are also crucial for understanding multilocal living. Here I shall limit my exposition on the first aspect, the interstices. The interstices – the in-between – has rarely been considered yet and it is a promising field of research and theory building for reflecting dynamics and movement as central aspects of multilocal living – also for the related topic of inequality.

Interstices

Most of the multilocalists I talked to so far commute between their places by train, followed by those who take the car. For the majority of them the time spent in transit is not experienced as lost time, especially for those travelling by train. The transit time has specific functions. Depending on the means of

---

6 In Summer 2008 a group of sociologists at the TU Chemnitz (Germany) finished their DFG-funded research on households, which for job reasons are organised in different living places. The study applies action-theoretical concepts and is based on qualitative interviews with adult household members. The scientists came up with a range of seven types of job-related multilocal households called 'Verschickung' (dispatch), 'Kolonisierung' (colonisation), 'Re-Zentrierung' (re-centralisation), 'Doppelleben' (double life), 'Bi-Polarisierung' (bi-polarisation), 'Expedition' (expedition), 'Drift' (Petzold et al. 2009, forthcoming).

transport, it may be a time to relax or sleep, to read or learn,[7] a time to have an intense and good time with (or without) his or her child in tow, a time to mentally prepare for the other (working or leisure) place or to reduce stress between work and home instead of *at* home (where others might suffer). For some it is just good to have a (more or less) quiet and private time between two places of expectations, e.g. workplace and family home. One multilocalist I met spends the time on the train to make notes for future conversations with his wife which had been truncated by their parting after the weekend. Although multilocalists cast around seek for strategies to benefit from the time in-between, commuting can also be stressful, depending on the specific circumstances.[8]

Here, interstices are not understood as something neutral between a point of origin and a destination (Kaufmann 2002), but as places created by imaginings and stories as well as by practices of *place making*[9] (Vonderau 2003). They constitute horizons of meaning for people living in and between different places (Rolshoven 2004, 215). Only when we consider *the in-between* can we overcome the obsolete dualism of the assured settledness on the one hand and the disturbing mobility on the other (Rolshoven 2005). Interstices generate distances, which separate and connect the different living places; they interrupt and convey (Koenen 2003, 156). According to Johanna Rolshoven, relevance-oriented science has to follow the mobile human individual to comprehend the new meanings of the interstices (Rolshoven 2004, 215). But the in-between does not merely exist on this physical level. In addition to interstices as imaginary places, one can find other (rather novel) kinds of interstices on a virtual level: the Internet (cyberspace) or today's mobile telephone networks are the most striking examples.

How do multilocalists cope with these interstices? How do they give meaning to them? One could think of certain activities and rituals like the coffee at the railway station, booting up the laptop after getting on the train or putting the sleeping mask on. One of the persons I interviewed, a musician, told me about three things he would do on arrival at a hotel room: close the curtains, choose the most comfortable room lighting and turn on CNN (or alternatively a local news station). Paradoxically he says: 'Then I know I am up to date, then I feel the city, what's going on, I get the groove' [translation by the author].

---

7 Some companies provide language lessons with a teacher who joins the customer on regular train rides, e.g. flyingteachers (www.flyingteachers.ch).

8 See Rolshoven and Winkler (2008, forthcoming) for another perspective on perceptions and experiences of commuters.

9 In her research on experiences of places in a mobile world, Asta Vonderau identifies *place making* as a 'dynamic place decision strategy' or, in other words, an ongoing process of positioning oneself towards different places (Vonderau 2003, 25).

How do multilocalists perceive the interstices, the in-between? This question is about explicit and implicit norms that exist in the interstices. Bänz Friedli, a journalist from Zurich who wrote a weekly train commuters' column in a free newspaper for several years, can be quoted as an example: more than once he complained about 'wrong commuters' (also the title of one of his columns). 'Wrong commuters' can be identified by their 'misconduct' on the train. One of the biggest mistakes one can make is to ask if a seat is already taken, because, he states, it is an implicit rule among commuters that no free seat is taken as yet (Friedli and Egger 2003, 97). As commuters have their rush hours, this rule is valid only temporarily.

And finally, how can the interrelations between *acting in space*, *perceiving space* and the *physical space* be characterised? Materialised space as a framework of dynamic appropriation processes can be supportive or obstructive (Rolshoven 2001). A typical example are functionally differentiated train compartments: first class, second class, business class, children's playground carriage, compartments for breast-feeding mothers etc. Today we have companies specialising in transit equipment (furniture) to increase the mobile customers' comfort. Markus Schröer (2006) identifies an increasing alignment of nomads and settlers expressed in two trends: making the means of transport homelike and the mobilisation of home living (Schröer 2006, 119–20).

In the effort to make transit space more comfortable, inequality among travellers is perpetuated and manifested.[10] For example, infrastructure in trains offers more and more options – but not for everyone. Not all trains have sockets to plug in laptops, mobile phone chargers etc., sometimes only first class carriages are equipped with them. The Swiss Federal Railway (SBB) has just started to provide wireless LAN, but only on a few major routes and only for first class passengers. These relatively new infrastructural developments and differentiations influence motives and behaviour of the (potential) customers greatly. It is a very dynamic process: while more and more people use their mobile phones on public transport, the railway companies are seeking solutions for those who do not want to be disturbed and are creating quiet areas (which, by the way, works very well in Switzerland, but not on Austrian trains).

Being able to use mobile phones and the Internet on trains also makes different interstices, i.e. concrete transit and virtual space overlap and/ or intersect. Nowadays (at least some) people can be on the road and surf through cyberspace at the same time. People acting in space while en route between their different living places shows that the space of movement has

---

10 In this context the dispute about the privatisation of public space and diverse exclusion mechanisms that come with it shall be mentioned. In his work about the interrelations between crime and the production of secure architecture, Michael Zingangel discusses the leading role of the German Federal Railways (DB) (Zinganel 2003) in developing exclusive semi-private spaces.

very important and individually specific functions of connecting different localities. In different dimensions we observe an individual competence of connecting different places via physical, cultural and virtual mobilities. This competence to be mobile on different levels and to conduct a satisfying life *because* or *despite* living multilocally is not equally distributed and is not likely to become so in the future. Therefore, it is a crucial dimension and source of social inequality.

Multilocal Living and Inequality: Interfaces

Sometimes I am asked what I do for a living. Being well aware that the answer should not take too long, I start with an introductory sentence and some examples of multilocal ways of life. Among the different possible reactions, often two contrasting and yet related issues are then raised – depending on the examples and which of them catches the attention of the person I am talking to. One of the reactions is: 'Only wealthy people can afford to keep two homes'. In an academic context, a similar observation is expressed in the argument that multilocal living is a phenomenon mainly affecting a highly educated elite – an idea that is reflected in the largely one-sided scientific attention for specific forms of multilocal living. Others commiserate with the 'Poor people who are forced to be mobile and away from home!' Such sympathy is based on the assumption that inhuman economic conditions make it impossible to live a 'normal' life. In academic texts the negative point of view can be summed up as: being mobile tends to be unhealthy and endangers family and other social life (e.g. Sennett 2000; Arbeiterkammer Wien 2005). Of course both objections are partly true for some forms of multilocal living and in some contexts, but the reality of multilocalists is rarely that one-dimensional.

As shown by the examples above, the phenomenon has entered societal mainstream. In the context of social inequality, the main issues arise along and between multilocal living as a privileged option and multilocal living as a constraint. The framework of a multilocal way of life does not provide the same set of opportunities and constraints to each multilocal person. In the same way the origins, the motives, the forms and possibilities in everyday life – in short, the living experience of the specific individual multilocal arrangement – can express utterly diverse dimensions of inequality.

Inequality means that access to generally available and desirable goods and positions, which include non-equal opportunities for power and interaction, is not equally distributed and therefore the life chances of the affected individuals, groups and societies are obstructed or facilitated (Kreckel 1992, 17). Multilocal living as such includes a series of aspects that are crucial when we talk about inequality: work, housing, social networks, travel, mobility and many others.

In the following paragraphs I shall discuss several less obviously materialised interfaces between the themes of multilocal living and social inequality.

First of all, there is no general answer to the question if and how multilocal living influences the structure of social inequality. From the variety of motives for and expressions of multilocal arrangements we can discover very different forms of that interaction. According to the theory of individualisation (Beck, 1986) we cannot characterise large social groups of multilocalists, which are connected along a certain structural relation. Of course not all multilocal arrangements are individualised to the same degree. As examples might demonstrate best, some are more limited by structural conditions than others. There is the worker from Leipzig who lives on a camp site in Basle from Monday to Friday. Each week he drives back home hundreds of kilometres to spend the weekends with his family in eastern Germany. A lack of labour market opportunities, combined with the wish to stay embedded in the home region, leads to a situation, which takes its toll on all family members. However, the many people from eastern Germany who take on jobs in western Germany, Switzerland or Austria these days while keeping their homes cannot be characterised as a coherent group concerned with the same specific factors of inequality. We have to differentiate them according to their motives, their values, their socio-demographic characteristics, their specific forms of organising the multilocal arrangements etc. An engineer from East Berlin, young and single, is faced with completely different conditions before and after deciding to move to Switzerland. Unlike his compatriot, he was offered a financially attractive job option allowing him to keep up his apartment in Berlin for regularly visiting friends and relatives. He is less restricted in his actions than the worker from Leipzig, he can pursue the pleasure principle better – at least to a certain degree. So one could assume that this young and mobile engineer is in some aspects of his multilocal lifestyle more like the retired couple from Switzerland spending the winter months on the Spanish Costa Blanca. If this was true, we could also assume that inequality as one aspect of living multilocally would not follow the obvious forms of classification (e.g. differentiations between leisure and work), but it would lie somewhere in-between, against the grain of common ideas of differentiating multilocal lifestyles. Further research will hopefully reveal more on this. The big challenge then will be to develop a typology or a model that manages to do justice to the high complexity of the phenomenon.

So far the collected empirical data leads to the following idea: every multilocal arrangement can be positioned on a continuum between the conflicting poles of freedom and constraint.[11] All actors make an active decision, at least in part, but the individual or collective (e.g. within a household) opportunities

---

11 The idea of describing multilocal living by using a continuum was inspired by Karin Jurcyk on the occasion of a project group meeting on multilocality research in Munich in December 2007.

differ considerably. Decisions in multilocal living are usually not made just with reference to the individual, i.e. the multilocalist him- or herself, but to a network of family and other social relations.

Social inequalities in living multilocally can also be related to the topic of rootlessness I mentioned earlier. A multilocal way of life does not necessarily lead to the loss of place-relatedness. Nevertheless there are forms of multilocal living that engender or include a feeling of loneliness – like the father with no job opportunities in his home town who misses his family during the workdays in the far away city – and even rootlessness: 'You're always travelling between worlds which are utterly different' [translation by the author], as one multilocalist pointed out, who is commuting between three job- and leisure-related places and holds down a stressful job. Thus multilocal living is not necessarily related to a mobile lifestyle, as Vincent Kaufmann emphasises for mobility in general: '... people can be very mobile without having a lifestyle based on fluidity, but one that is based on the will [or the constraints] to separate their lives socially and spatially into distinct areas' (Kaufmann 2002, 58–9).

The way a multilocal arrangement is construed also influences the acceptance of a certain form of multilocal living. What Vincent Kaufmann observes for mobility forms in general (Kaufmann 2002, 102) can be said for multilocal ways of life. There are some dominant forms which are valued and relatively easy to realise and can be lived unobtrusively. However, multilocalists who do not comply with generally accepted models are often forced to explain and justify themselves because there is something 'fishy' in their lives. Thus the permanent campers I interviewed in Basle reported that they are sometimes called 'gypsies'. In fact, as a camper club they changed the camping rules to get rid of those travellers who had come by from time to time.

Out of these reflections based on my empirical data I put forward the following hypotheses on the interaction between multilocal living and inequality:

- Multilocal living as a specific organisation in time and space is a highly relevant factor for new inequalities, because it obviously reflects the end of certain traditional bonds, while at the same time it comes out as a new restricting dimension of inequality. Vincent Kaufmann (2002, 101) emphasises that we are confronted with 'new factors of social differentiation that are no longer built around defined territorial limits, but instead around space and time'.
- The new characteristics of multilocal living as a factor of inequality do not primarily follow the common structural dimensions, but are transverse to them.
- Multilocal living can be a strategy to maintain the existing status of an individual.
- Multilocal living can be a strategy to upgrade one's social status.

- Multilocal living can be a way to balance inequalities between economically privileged and non-privileged regions (e.g. the western and eastern parts of Germany), with individuals and households acting out structural change (Petzold et al. 2009, forthcoming).
- The fact, the experience and perception of multilocal living as a factor of inequality depends on the dimension of acceptable alternatives.
- The fact, experience and perception of multilocal living as a factor of inequality depends on how much this specific lifestyle agrees with personal values and ideas of a 'good life'.
- The different multilocal lifestyles differ according to the degree of social acceptance in their social environment.

Further research should go into the degree of social acceptance or discrimination of different multilocal lifestyles. An analysis of the motives and strategies of different multilocalists can provide new insights into the relationship of multilocal living and social change, as well as considering (potential) emerging new social inequalities. In this discussion the lack of terms to describe these new life realities also raises problems. Doubtlessly, we need a commonly agreed vocabulary to explain and understand such phenomena.

New (Mobile) Lifestyles – New Terms and Concepts?

The phenomenon of multilocal living not only lacks scientific research and terms but also a common parlance in which to denote it accurately. This state of affairs became clear to me when I was talking to multilocalists. I interviewed a musician who had lived in and between Switzerland and Los Angeles for quite a few years and had now decided to stay in the United States for some time to pursue his career, but who still kept his apartment in Switzerland. He is often asked when he will emigrate. But he totally dislikes the term *emigration*. In his understanding it is much too final and static, he claimed. There are no adequate expressions to describe these people's life realities. Also for me it is sometimes difficult to use the terms appropriately. I sometimes confused multilocalists when I used the term *commuting*. For them commuting is usually understood as daily commuting. Another example is the term *second home*. Empirical evidence shows that this popular expression is rather vague because the hierarchy of meanings of the different places is more complicated. It does not reflect the actual usage of more than one home and it is not the expression people would use for their own multilocal setting. This is especially true of permanent campers or people who move seasonally, e.g. between Switzerland and Spain. They attribute the same value to the different places and feel attached to both as a place called 'home'.[12] Thus, in future research, it would be rewarding to find out which terms and concepts about their way of life the multilocalists themselves provide.

Conclusion

Multilocal living in its various forms and interpretations is a highly relevant phenomenon in late-modern societies. It occurs in a wealth of variants. Multilocal living is a framework for and life practice of people who integrate different places with specific functions and meanings into a – for them – meaningful whole. The topic of this chapter, multilocal living in the sense of having and using more than one home, is seen as a specific form of multilocality. The analysis of multilocal living reveals important dimensions of social change. In this chapter I have pointed out some generic interrelations between multilocal lifestyles, space and inequality. Multilocal living is highly relevant for all kinds of spatial questions as well as an important aspect in the context of social inequality. The dimensions of multilocal living, mobility, space and social inequality should be thought of as interdependent variables that influence (and often amplify) each other on different levels. This is not just a theoretical insight but something that crucially affects the everyday life of people who are involved in multilocal arrangements in one way or another. Thus, it is vital to consider all these aspects in upcoming research on this issue. Multilocal living is a broad phenomenon of growing importance that has to be studied in an inter- and transdisciplinary setting. Joint efforts of experts in the social and cultural sciences (such as sociology, cultural anthropology or geography), transportation sciences and planning disciplines can help to shed some light on the new old phenomenon of multilocal living and its implications for today's societies in a holistic and hence innovative perspective.

References

Albrow, M. (1997), 'Auf Reisen jenseits der Heimat. Soziale Landschaften in der Stadt', in Beck, U. (ed.), *Kinder der Freiheit* (Frankfurt am Main: Suhrkamp), 288–314.

---

12 Alternatively to *second home* the term *l'autre maison* (the other house) is used, meaning a space imbued with possibilities which allows people to 'double' their lifestyles (Dubost 1998; Bonnin and de Villanova 1999). German Cabaret Artist Gerhard Polt, who lives in three different places in Germany and Italy, expresses his personal multilocal experience in one sentence: 'If you ask me where my home is, I have to ask back: When?' (broadcast on TV channel Bayern 3, 5 May 2007 in the programme 'Poltrait zum 65. Geburtstag' by Ute Casper).

Appadurai, A. (1991), 'Global Ethnoscapes. Notes and Queries for a Transnational Anthropology', in Fox, R.G. (ed.), *Recapturing Anthropology. Working in the Present* (Santa Fé, NM: School of American Research Press), 191–238.

Arbeiterkammer Wien (2005), 'Überfordert vom Arbeitsweg? Was Stress und Ärger am Weg zur Arbeit bewirken können', *Arbeiterkammer Wien* <http://wien.arbeiterkammer.at/pictures/d36/Verkehr_Arbeitsweg_Studie2006.pdf>, accessed 10 August 2007.

Augé, M. (1994), *Orte und Nicht-Orte. Vorüberlegungen zu einer Ethnologie der Einsamkeit* (Frankfurt am Main: S. Fischer).

Axhausen, K.W. (2007), 'Activity Spaces, Biographies, Social Networks and their Welfare Gains and Externalities: Some Hypotheses and Empirical Results', *Mobilities* 2:1, 15–36.

Beck, U. (ed.) (1997), *Kinder der Freiheit* (Frankfurt am Main: Suhrkamp).

Beck, U. (ed.) (1998), *Perspektiven der Weltgesellschaft* (Frankfurt am Main: Suhrkamp).

Beck, U. (ed.) (1997), *Was ist Globalisierung? Irrtümer des Globalismus – Antworten auf Globalisierung* (Frankfurt am Main: Suhrkamp).

Beck, U. (1986), *Risikogesellschaft. Auf dem Weg in eine andere Moderne* (Frankfurt am Main: Suhrkamp).

Beck, U. (2007), *Weltrisikogesellschaft. Auf der Suche nach der verlorenen Sicherheit* (Frankfurt am Main: Suhrkamp).

Beck, U. and Lau, C. (eds) (2004), *Entgrenzung und Entscheidung. Perspektiven reflexiver Modernisierung* (Frankfurt am Main: Suhrkamp).

Becker, S., Bimmer, A.C., Braun, K., Buchner-Fuhs, J., Gieske, S. and Köhle-Hezinger, C. (eds) (2001), *Volkskundliche Tableaus* (Münster: Waxmann).

Beer, B. (ed.) (2003), *Methoden und Techniken der Feldforschung* (Berlin: Dietrich Reimer).

Bonnin, P. and de Villanova, R. (eds) (1999), *D'une maison l'autre. Parcours et mobilités résidentielles* (Grane: Éditions Créaphis).

Bonss, W., Kesselring, S. and Weiss, A. (2004), "Society on the Move'. Mobilitätspioniere in der Zweiten Moderne', in Beck, U. and Lau, C. (eds), *Entgrenzung und Entscheidung. Perspektiven reflexiver Modernisierung* (Frankfurt am Main: Suhrkamp), 258–80.

Bonss, W. and Kesselring, S. (1999), 'Mobilität und Moderne. Zur gesellschaftstheoretischen Verortung des Mobilitätsbegriffs', in Tully, C. (ed.), *Erziehung zur Mobilität. Jugendliche in der automobilen Gesellschaft* (Frankfurt am Main: Campus).

Breckner, I. (2002), '"Wohnen und Wandern' in nachindustriellen Gesellschaften', in Döllmann, P. and Temel, R. (eds), *Lebenslandschaften. Zukünftiges Wohnen im Schnittpunkt von privat und öffentlich* (Frankfurt am Main: Campus), 145–53.

Collmer, S. and Kümmel, G. (2005), *Ein Job wie jeder andere? Zum Selbst- und Berufsverständnis von Soldaten* (Baden-Baden: Nomos).

Döllmann, P. and Temel, R. (eds) (2002), *Lebenslandschaften. Zukünftiges Wohnen im Schnittpunkt von privat und öffentlich* (Frankfurt am Main: Campus).
Dubost, F. (ed.) (1998), *L'autre maison. La 'résidence secondaire', refuge des generations* (Paris: Autrement).
Englisch, G. (2001), *Jobnomaden. Wie wir arbeiten, leben und lieben werden* (Frankfurt am Main: Campus).
Fox, R.G. (ed.) (1991), *Recapturing Anthropology. Working in the Present* (Santa Fé, NM: School of American Research Press).
Friedli, B. and Egger, A. (2003), *Ich pendle, also bin ich* (Zürich: Hagenbuch).
Froschauer, U. and Lueger, M. (2003), *Das qualitative Interview. Zur Praxis interpretativer Analyse sozialer Systeme* (Wien: WUV Universitätsverlag).
Fuhrer, U. and Kaiser, F.G. (1994), *Multilokales Wohnen. Psychologische Aspekte der Freizeitmobilität* (Bern: Huber).
Gallent, N. and Tewdwr-Jones, M. (2000), *Rural Second Homes in Europe. Examining Housing Supply and Planning Control* (Aldershot: Ashgate).
Gebhardt, W. and Hitzler, R. (eds) (2006), *Nomaden, Flaneure, Vagabunden. Wissensformen und Denkstile der Gegenwart* (Wiesbaden: VS Verlag für Sozialwissenschaften).
Giordano, C. and Rolshoven, J. (eds) (1999), *Europäische Ethnologie – Ethnologie Europas* (Fribourg: Universitätsverlag).
Glaser, B.G. and Strauss, A.L. (1968), *The Discovery of Grounded Theory. Strategies for Qualitative Research* (London: Weidenfeld & Nicolson).
Guzzoni, U. (1999), *Wohnen und Wandern* (Düsseldorf: Parerga).
Gyr, U. and Rolshoven, J. (eds) (2004), *Zweitwohnsitze und kulturelle Mobilität. Feldforschungsberichte* (Zürich: Volkskundliches Seminar).
Hall, M.C. and Müller, D.K. (2004), *Tourism, Mobility and Second Homes. Between Elite Landscape and Common Ground* (Clevedon: Channel View Publications).
Hannerz, U. (1996), *Transnational Connections. Culture, People, Places* (London: Routledge).
Jürges, H. (1998), 'Beruflich bedingte Umzüge von Doppelverdienern. Eine empirische Analyse mit Daten des SOEP', *Zeitschrift für Soziologie* 27:5, 358–77.
Kalter, F. (1994), 'Pendeln statt Migration? Die Wahl und Stabilität von Wohnort-Arbeitsort-Kombinationen', *Zeitschrift für Soziologie* 23:6, 460–76.
Kaufmann, V. (2002), *Re-thinking Mobility. Contemporary Sociology* (Aldershot: Ashgate).
Kaufmann, V., Bergman, M.M. and Joye, D. (2004), 'Motility: Mobility as Capital', *International Journal of Urban and Regional Research* 28:4, 745–65.
Kaufmann, V., Kesselring, S., Manderscheid, K. and Sager, F. (2007), 'Mobility, Space and Social Inequality', *Swiss Journal of Sociology* 33:1, 5–6.

Kecskes, R. (1994), 'Abwanderung, Widerspruch, Passivität. Oder: Wer zieht wann um?', *Zeitschrift für Soziologie* 23:2, 129–44.
Koenen, E.J. (2003), 'Öffentliche Zwischenräume. Zur Zivilisierung räumlicher Distanzen', in Krämer-Badoni, T. and Kuhn, K. (eds), *Die Gesellschaft und ihr Raum. Raum als Gegenstand der Soziologie* (Opladen: Leske + Budrich), 155–72.
Krämer-Badoni, T. and Kuhn, K. (eds) (2003), *Die Gesellschaft und ihr Raum. Raum als Gegenstand der Soziologie* (Opladen: Leske + Budrich).
Kreckel, R. (1992), *Politische Soziologie der sozialen Ungleichheit* (Frankfurt am Main: Campus).
Kreutzer, F. and Roth, S. (eds) (2006), *Transnationale Karrieren, Biografien, Lebensführungen und Mobilität* (Wiesbaden: VS Verlag für Sozialwissenschaften).
Löfgren, O. (1995), 'Leben im Transit? Identitäten und Territorialitäten in historischer Perspektive', *Historische Anthropologie* 3:3, 349–63.
Löfgren, O. (1999), *On Holiday. A History of Vacationing* (London: University of California Press).
Luka, N. (2006), Placing the 'Natural' Edges of Metropolitan Region through Multiple Residency: Landscape and Urban Form in Toronto's 'Cottage Country", unpublished PhD thesis.
Marcus, G.E. (1995), 'Ethnography in/of the World System. The Emergence of Multi-Sited Ethnography', *Annual Review of Anthropology* 24:1, 95–117.
Meier, B. (1999), *Mobile Wirtschaft – immobile Gesellschaft. Gesellschaftspolitische Konsequenzen der Globalisierung* (Köln: Deutscher Instituts-Verlag).
Merkel, I. (2003), 'Außerhalb von Mittendrin. Individuum und Kultur in der zweiten Moderne', *Kulturation. Online-Journal für Kultur, Wissenschaft und Politik* 1, 1–20.
Opaschowski, H.W. (1999), *Umwelt, Freizeit, Tourismus* (Opladen: Leske + Budrich).
Pelizäus-Hoffmeister, H. (2001), *Mobilität: Chance oder Risiko? Soziale Netzwerke unter den Bedingungen räumlicher Mobilität – das Beispiel freie JournalistInnen* (Opladen: Leske + Budrich).
Petzold, K. et al. (2009), 'Multilokale Haushaltstypen', *Informationen zur Raumentwicklung* (forthcoming).
Rolshoven, J. (2001), 'Gehen in der Stadt' in Becker, S., Bimmer, A.C., Braun, K., Buchner-Fuhs, J., Gieske, S. and Köhle-Hezinger, C. (eds), *Volkskundliche Tableaus* (Münster: Waxmann), 11–27.
Rolshoven, J. (2002a), 'Depopulation and Reterritorialisation in Peripheral Regions: New Social Spaces in the South of France', <http://www.ipk.unizh.ch/studium/download/rolshoven_mobility.pdf>, accessed 28 May 2008.
Rolshoven, J. (2002b), 'Südliche Zweitwohnsitze. Ein Beitrag zur kulturwissenschaftlichen Mobilitätsforschung', *Schweizerisches Archiv für Volkskunde* 98:2, 345–56.

Rolshoven, J. (ed.) (2003a), *Hexen, Wiedergänger, Sans-Papiers* ... (Marburg: Jonas).
Rolshoven, J. (2003b), 'Von der Kulturraum- zur Raumkulturforschung. Theoretische Herausforderungen an eine Kultur- und Sozialwissenschaft des Alltags', *Zeitschrift für Volkskunde* 99:2, 189–213.
Rolshoven, J. (2004), 'Mobilität und Multilokalität als moderne Alltagspraxen. Ethnographien kultureller Mobilität', in Gyr, U. and Rolshoven, J. (eds) (2004), *Zweitwohnsitze und kulturelle Mobilität. Feldforschungsberichte* (Zürich: Volkskundliches Seminar).
Rolshoven, J. (2005), 'Region und Regionalkultur. Räume mobiler Alltagskulturen', presented at an assembly of the Schweizerische Gesellschaft für Kulturwissenschaft, Zurich, 6 June, <http://www.ipk.unizh.ch/studium/download/rolshoven_mobility.pdf>, accessed 28 May 2008.
Rolshoven, J. (2006), 'Woanders daheim. Kulturwissenschaftliche Ansätze zur multilokalen Lebensweise in der Spätmoderne', *Zeitschrift für Volkskunde* 2:3, 179–94.
Rolshoven, J. (2007), 'Multilokalität als Lebensweise in der Spätmoderne', *Schweizerisches Archiv für Volkskunde* 103:2, 157–79.
Rolshoven, J. and Winkler, J. (2008), 'Auf den Spuren der schönen Pendlerin', *kuckuck. Notizen zur Alltagskultur* (forthcoming).
Römhild, R. (2003), 'Practised Imagination. Tracing Transnational Networks in Crete and Beyond', *AJEC* 11:2, 159–90.
Schiffauer, W. (1991), *Die Migranten aus Subay. Türken in Deutschland. Eine Ethnographie* (Stuttgart: Klett-Cotta).
Schlehe, J. (2003), 'Formen qualitativer ethnographischer Interviews', in Beer, B. (ed.), *Methoden und Techniken der Feldforschung* (Berlin: Reiner), 71–93.
Schneider, N.F., Limmer, R. and Ruckdeschel, K. (2002a), *Berufsmobilität und Lebensform. Sind berufliche Mobilitätserfordernisse in Zeiten der Globalisierung noch mit Familie vereinbar?* (Stuttgart: Kohlhammer).
Schneider, N.F., Limmer, R. and Ruckdeschel, K. (2002b), *Mobil, flexibel, gebunden. Familie und Beruf in der mobilen Gesellschaft* (Frankfurt am Main: Campus).
Schröer, M. (2006), 'Mobilität ohne Grenzen? Vom Dasein als Nomade und der Zukunft der Sesshaftigkeit', in Gebhardt, W. and Hitzler, R. (eds), *Nomaden, Flaneure, Vagabunden. Wissensformen und Denkstile der Gegenwart* (Wiesbaden: VS Verlag für Sozialwissenschaften), 115–25.
Selle, G. (2002), 'Innen und außen – Wohnen als Daseinsentwurf zwischen Einschließung und erzwungener Öffnung', in Döllmann, P. and Temel, R. (eds), *Lebenslandschaften. Zukünftiges Wohnen im Schnittpunkt von privat und öffentlich* (Frankfurt am Main: Campus).
Sennett, R. (2000[1998]), *Der flexible Mensch. Die Kultur des neuen Kapitalismus* (München: Goldmann).
Stock, M. (2006), 'L'hypothèse de l'habiter poly-topique: pratiquer les lieux géographiques dans les sociétés à individus mobiles', *EspacesTemps.net*

<http://www.espacestemps.net/document1853.html>, accessed 5 May 2008.
Tully, C.J. and Baier, D. (2006), *Mobiler Alltag. Mobilität zwischen Option und Zwang – Vom Zusammenspiel biographischer Motive und sozialer Vorgaben* (Wiesbaden: VS Verlag für Sozialwissenschaften).
Tully, C.J. (ed.) (1999), *Erziehung zur Mobilität. Jugendliche in der automobilen Gesellschaft* (Frankfurt am Main: Campus).
Urry, J. (1990), *The Tourist Gaze. Leisure and Travel in Contemporary Societies* (London: Sage).
Vonderau, A. (2003), *Geographie sozialer Beziehungen. Ortserfahrungen in der mobilen Welt* (Münster: LIT Verlag).
Wallace, C., Sidorenko, E. and Chmouliar, O. (1999), 'The Central European Buffer Zone', in Giordano, C. and Rolshoven, J. (eds), *Europäische Ethnologie* (Fribourg: Universitätsverlag), 123–69.
Welz, G. (1998), 'Moving Targets. Feldforschung unter Mobilitätsdruck', *Zeitschrift für Volkskunde* 94:2, 177–94.
Werbner, P. (1999), 'Global Pathways. Working Class Cosmopolitans and the Creation of Transnational Ethnic Worlds', *Social Anthropology* 7:1, 17–35.
Wicker, H.-R. (1996), 'Flexible Cultures, Hybrid Identities and Reflexive Capital', *AJEC* 5:7, 7–30.
Widmer, M. (2003), 'Der Zwischenraum als Lebenswelt. "Sans-Papiers" in der Schweiz', in Rolshoven (ed.), *Hexen, Wiedergänger, Sans-Papiers ...* (Marburg: Jonas), 50–65.
Xiao Di, Z. and McArdle, N. (2001), 'Second Homes: What, How Many, Where and Who', *Joint Center for Housing Studies*, Harvard University <http://www.jchs.harvard.edu/publications/homeownership/di_n01-2.pdf>, accessed 3 October 2003.
Yeoh, B.S.A., Charney, M.W. and Tong, C.K. (eds) (2003), *Approaching Transnationalism. Studies on Transnational Societies, Multicultural Contacts and Imaginings of Home* (Boston: Kluwer Academic).
Zinganel, M. (2003), *Real Crime. Architektur, Stadt & Verbrechen. Zur Produktivkraft des Verbrechens für die Entwicklung von Sicherheitstechnik, Architektur und Stadtplanung* (Wien: Edition selene).
Zvonkovic, A.M., Richards, C., Humble, A. and Manoogian, M. (2005), 'Family Work and Relationships: Lessons from Families of Men whose Jobs Require Travel', *Family Relations* 54:3, 411–22.

Chapter 8
Class Divides within Transnationalisation – The German Population and its Cross-Border Practices

Steffen Mau and Jan Mewes

Introduction

In the last decades, sociologists put tremendous effort into demonstrating and explaining the ever increasing global interconnectedness. In this context, 'globalisation' became a key word as well as a key sociological concept of the late twentieth century, referring to 'the intensification of worldwide social relations which link distant localities in such a way that local happenings are shaped by events occurring many miles away and vice versa' (Giddens 1990, 64). According to this reading, everyone is exposed to this process, be it voluntarily or not. The same applies to individual mobility: 'All of us are, willy-nilly, by design or by default, on the move. We are on the move even if, physically, we stay put: immobility is not a realistic option in a world of permanent change' (Bauman 1998, 2). However, theoretical concepts that do not distinguish between an active way of globalisation (that is, for example, cross-border interaction and mobility) and a passive one, which just means to *experience* global interconnectedness, do blur the question of who the 'agents' of globalisation really are. Due to the dominance of research focused on macrostructural change, the role of microsociological actors within the processes of denationalisation both remains untheorised and empirically unexplored to a high degree (but see de Swaan 1995; Hannerz 1996; Nowicka 2006).

In this chapter, therefore, we take up the recently increased efforts to focus research on the 'human face of global mobility' (Favell et al. 2006). Both Social Network Analysis as well as the sociology of inequality provide telling arguments and examples for class-specific forms of cross-border mobility and interaction. Is it primarily or even solely the economic and social elites (Sklair 1991; Carroll and Fennema, 2002; Beaverstock, 2005) whose dealings in transnational fields of action and communication make the visions of a 'global society' (Albrow 1998) or 'world society' (Heintz 1982) a tangible prospect? Or can we observe an increase in cross-border mobility and a spatial extension of social relations equally in *all* strata of the population? We thus ask, through the lens of sociology of social inequality: to what extent is the participation in transnational fields of communication and interaction vertically stratified, for example by education and occupational status?

In order to find an answer to this question, we make use of the concept of *transnationalisation*.[1] This has several advantages over the use of standard concepts of globalisation: firstly, theorems of transnationalisation focus very centrally on the everyday individual level and secondly, they take greater account of empirical findings that also enable the demonstration of generative mechanisms of individual involvement in cross-border interactions. The term 'globalisation' is also unsuitable in many respects for describing the spatial extension of horizons of social action in an adequate manner. For example, scholars have shown that the spatial extension of cross-border transactions supports a trend towards 'OECD-isation' rather than *globalisation* (Beisheim et al. 1999; Zürn 1998). The concept of transnationalisation takes conscious account of these findings by investigating the spatial dimensions of cross-border processes. Simultaneously, this perspective distances itself from that of globalisation to such an extent that it does not necessarily regard the nation state as an 'outdated model' destined for extinction. Analytically, one can distinguish between the perspectives of 'transnationalisation from above' and 'transnationalisation from below' (Guarnizo and Smith 1998). Whereas the former perspective makes reference to cross-border interaction by corporative actors, the latter centres on the study of the direct social relationships between people living in different states (Vobruba 1995, 339; see also Mau 2007). It is this perspective that we take in our chapter .

We start by presenting arguments as to why there are likely to be clear class differences with regard to involvement and participation in transnational processes. This theoretical discussion develops the hypothesis that it is primarily groups with comparatively high educational capital and high occupational status that are involved in transnational fields of action and communication in their private lives. According to this assumption, transnationalisation is a highly unequal process, with transnationally oriented elites on the one hand and a greater number of lower status groups with strong national involvement on the other hand. To test the hypothesis of unequal transnationalisation, the empirical part of the chapter makes use of survey data gathered in the spring of 2006 as part of a representative survey of 2,700 German citizens, funded by the German Research Foundation (*Deutsche Forschungsgemeinschaft*).

---

1   In this chapter we refer to studies which are most often embraced by the term 'transnationalism'. As experts in that particular research field themselves raised the objection that the original term fails to comprise the structural *and* processual character of the phenomenon (Szanton Blanc et al. 1995, 684; Pries 2002, 264), we use the term 'transnationalisation' instead.

Unequal Transnationalisation?

The issue of social inequality of participation in new forms of border crossing is not self-explanatory. From a system-theoretical perspective, a world society is already regarded as realised as any person can potentially be an addressee of global communication (Luhmann 1997). This does not mean that everyone necessarily has to get in contact with everyone else in a world society, but at most that there is no longer any reason why interactions should end at the borders of the nation state. The world constitutes a unit of all possible communications (Greve and Heintz 2005). With reference to real networks and mobility we can also diagnose that the conditions for interaction and communication have changed to such a drastic extent that the 'world' is entering into social relations (Stichweh 2000, 233). For instance, structural changes and the increasing permeability of borders mean that people are coming into contact with more and more different persons, and that their personal networks are rapidly expanding. Simultaneously, the frequency of physical border-crossing activities is increasing in the form of short- and long-term mobility.

However, a closer look at real networks and cross-border mobility shows that the hypothesis of a global or transnational mobility for the entire population is misleading. Although we can assume that the majority of individual actors are indirectly involved in the field of interaction within the global society (as suggested by the research on the Small World phenomenon, see Watts 1999, 2003), this space only constitutes a subjectively relevant horizon of action for a limited number of individuals (Heintz 1976; see also Merton, 1995). The literature on globalisation and transnationalisation highlights the outstanding role of economical, political and social elites. Despite the many findings in the field of migration research on the emergence of transnational contacts and networks, it is still widely assumed that globalisation is ultimately brought forth and borne by elites (Bauman 1998, 2–3; Sklair 2001; Nowicka 2006, 18–19).

The evolution of global marketplaces, information and investment streams, and processes of international division of labour are accompanied, for example, by an internationalisation of many companies' operations – and along with the operations, the activities of their management staff. It is the class of the transnational experts that has the greatest benefits to anticipate from the spatial extension of social arenas: 'Economic interests motivate the global players to invest in emotional ties beyond cultural and linguistic barriers. In post-Fordist society, they have access to all the technical means for building and maintaining their social networks that also enable economic globalisation' (Brinkschröder 1999, 225 [authors' translation]).[2] As a result, changes have

---

2 Original German text: 'Die ökonomischen Interessen motivieren die *global players*, auch über kulturelle und sprachliche Barrieren hinweg in emotionale Bindungen zu investieren. Für den Aufbau und die Pflege ihrer sozialen Netzwerke

come about in the skills and qualifications required of the class of economic experts. For managers with international scope, transnational competence is now part of the basic toolkit. This refers to the ability to understand other codes, conventions, attitudes and modes of behaviour and to communicate and cooperate across cultural boundaries (Koehn and Rosenau 2002).

Although the focus is frequently on the economic decision-makers, the concept of the transnational expert class goes much further than this, including numerous occupations and activities in the political sector, in the administration or in knowledge-based organisations. Thus, alongside business actors and experts, political elites, for example, play a key role in the establishment of transnational political practices (Sklair 1991). The political sphere participates actively in the creation, maintenance and structuring of international regimes. Even bureaucracies, traditionally the stronghold of nation states' gatekeeping activities, are now part of networks of information exchange and mutual reporting and attention, because the system of regulations within the nation state is increasingly intertwined on an international or supranational level, creating greater need for coordination. Many specific committees and administrations, be they in the field of telecommunications, security, pensions or taxes, have to consult with experts from other countries in order to judge the legal and other implications arising from these intertwining circumstances correctly. Particularly within the European Union, we can find numerous direct interactions between the representatives of national administrative bodies and the European Commission or the representatives of other countries.

Also the third sector has been subject to great change. There is now a whole spectrum of NGOs with the word *international* in their very names (for example Amnesty International), implying that neither their membership nor their political aims are limited by national borders. Furthermore, the internationalisation of science and of the universities has now reached a very advanced stage. In accordance with the type of academic knowledge production, the communication networks of the sciences stretch far beyond the bounds of the nation state. One indicator, for example, is the increasing number of multinational authorships of academic articles (see Gerhards and Rössel 1999; Stichweh 1999). It is now perfectly normal for academics to (have to) enter into international communicative relationships in order to stay on top of their discipline (Stichweh 2000).

The new communication media ease and support the creation of transnational social networks, making the exchange of information fast and simple. Nevertheless, face-to-face interactions still play an immanently important role in forming and maintaining long-distance relationships (Urry 2004, 2007). We can therefore assume that there is a relationship of mutual intensification between mobility and spatially extended network relationships,

---

stehen ihnen im Postfordismus alle technischen Mittel zur Verfügung, die auch die ökonomische Globalisierung ermöglichen'.

as mobility increases the number and diversification of possible interaction partners and enables social relationships to be maintained through direct contact. Mobility can thus be regarded as a creative nexus and as a 'tool to maintain social networks' (Nowicka 2006, 37).

Simultaneously, transnational social networks that develop in a professional setting have the advantage that the at times expensive and time-intensive private planning of the occasional meetings essential for maintaining social ties is not necessary, as the wealth of events jointly visited by all those concerned often offers sufficient opportunities for face-to-face communication. The international networking of experts thus has a doubly positive effect on their international mobility, firstly in a direct sense – through participation in periodical 'occasioned meetings' – and secondly indirectly – through the establishment of cross-border friendships.

However, it is not only the specific profiles of occupational mobility that involve certain groups in cross-border interactions. As a whole, we can also assume unequal participation in transnational activities along the lines of social structure because the 'transnational competences' (Koehn and Rosenau 2002), that is linguistic, cultural, social and cognitive skills, are unequally distributed. One group generally ascribed with special skills for cross-border activities is the highly educated. It is said that they are typically far less restricted by the borders of the nation state, due to a greater openness towards other cultures, knowledge of foreign languages and a frame of reference reaching beyond the local level (Konrad 1984, 209; Hannerz 1996, 102ff; see also Mau et al. 2008).

We know, from research into social networks, that the majority of personal relationships of those in the educationally poor social strata are strongly anchored on the local level and dominated by family contacts (Allan 1979, 1989). Neighbourhood-ties also have a greater significance in these strata than in higher-status occupational groups (see MacDonald et al. 2005). In contrast, the 'personal communities' (Wellman et al. 1988) of more highly educated groups are characterised by a greater spatial range, greater heterogeneity and a greater accentuation of ties based on friendship or acquaintances (Marsden 1987; Willmott 1987; Allan 1998; Ohnmacht et al. 2008; Axhausen and Frei 2007). Part of the social inequality in the structure of social networks can be explained by a class-specific 'motility' (Kaufmann 2002), that is the individuals' very *options* and *opportunities* to be mobile tend to vary across social classes (see for example Cass et al. 2005).The lower classes appear to be severely constrained as regards their opportunities to maintain long-distance relations, as the costs of physical mobility, be it by means of cars or by public transport, still prevent them from making long-distance journeys for many a time (albeit it must be acknowledged that the general costs of mobility haven sunken during the last decades, see Axhausen 2007). In this regard, Larsen et al. (2006: 54) refer to clear 'mobility divides' in Western societies. In the process of transnationalisation, which entails rather complex and intercultural conditions of interaction, these class-specific differences with regard to the range of social networks are presumably becoming even more marked.

It is far easier for those with higher social status to accumulate transnational social capital and to cope with the demands connected with these networks. Highly educated individuals and persons in higher occupational positions, therefore, may tend to be more spatially mobile, which increases the chances of making more contacts with an increasing number of people. Fundamentally, we assume that the contacts and network relationships formed in the professional setting also have an effect on private forms of networking (see Wittel 2001, 69; Kennedy 2004; Völker and Flap 2007) and that group differences in working life also extend into private life.[3]

From the perspective of social structure, the above considerations lead to the thesis that it is particularly vertical strata categories that should manifest the differing grades of transnationalisation. Due to their social positions, their skills and their 'motility', groups with high educational capital and comparatively high occupational status may form that segment of the population that takes on a pioneering role in the process of denationalisation of horizons of social action. They can be ascribed with both an increased interest in transgressing and overcoming the borders of the nation state and the social and cultural tools required for cross-border communication and action on a successful and socially satisfactory basis. We can assume a close connection between occupational positions and private forms of transnational activities, as contacts are made and transnational skills are gained in working contexts, which also play a significant role in private life.

Empirical Findings of Unequal Transnationalisation

To subject our hypothesis of unequal transnationalisation to empirical examination, we make use of data from the *Survey Transnationalisierung 2006*. This representative survey of the German population was planned as part of the project 'Transnationalisation of Social Relationships' at the University of Bremen, funded by the German Science Foundation (*Deutsche Forschungsgemeinschaft*), and carried out in cooperation with the social science research institute IPSOS. A total of 2,700 people were interviewed by

---

3 According to some more pessimistic representatives of network studies, the chance of making many contacts through frequent changes of location is, however, linked with the risk that the social relationships of mobile individuals tend to have a shorter lifespan. Boissevain (1974, 87) even uses the saying 'out of sight, out of mind' with reference to the social networks of geographically mobile individuals. In contrast, Larsen et al. (2006) find in their study that people attach the same importance to friends and relatives living abroad than they do with regard to those living within the country.

telephone (CATI design) on various aspects of their personal integration into transnational interaction and activities. The population of the study consists of all German-speaking persons from the age of 16 with German nationality,[4] who live in private households in the Federal Republic of Germany and can be contacted by a telephone landline.[5]

Is there a correlation between unequal social status positions and involvement in transnational spaces of action and communication? We start our analyses by descriptively investigating how involvement in transnational interaction and different forms of cross-border mobility is linked to educational attainment on the one hand and to occupational status on the other hand. Subsequently, a multivariate regression will clarify to what extent differing grades of transnationalisation is determined by sociostructural characteristics.

For scrutinising the link between educational level and the degree of transnationalisation, we distinguish three educational groups, namely those with higher education entrance qualifications (high educational level), those with intermediate secondary school-leaving certificate (average educational level) and those with a lower secondary school-leaving certificate or no certificate (low educational level). Corresponding with our hypotheses, the involvement in transnational networks and cross-border mobility increases very sharply with education. The differences between the two groups are very clear in all categories included in the survey (Figure 8.1). Almost 70 per cent of respondents with higher educational levels have regular private contact to a person living abroad, whereas the corresponding figure for persons with lower educational levels is just under 40 per cent. The differences between contacts to non-Germans living outside of Germany are particularly stark. In this case it is non-family ties in particular, that is, contact to friends and acquaintances, where education-specific differences are most marked. We can assume that these ties are very strongly determined by individual preferences and efforts towards initiating and maintaining them. The orientation towards such network relationships is four times as high among the higher educated than among those with lower education.

We also find great class differences in the geographical spread of these transnational networks. We know that German people's transnational networks are not globally distributed, but are concentrated on a certain geographical area. Approximately 85 per cent of the contacts abroad established in the survey are concentrated in only 20 of the 193 world states

---

4   This includes people with dual citizenship, provided one of their citizenships is German.

5   As sample outcomes are not equally distributed, we have used a weighting factor for the findings presented below, which adjusts the unweighted sample structure to the official statistics. To do so, a weighting was undertaken on the basis of the characteristics age, gender, *Bundesland*, community size and education.

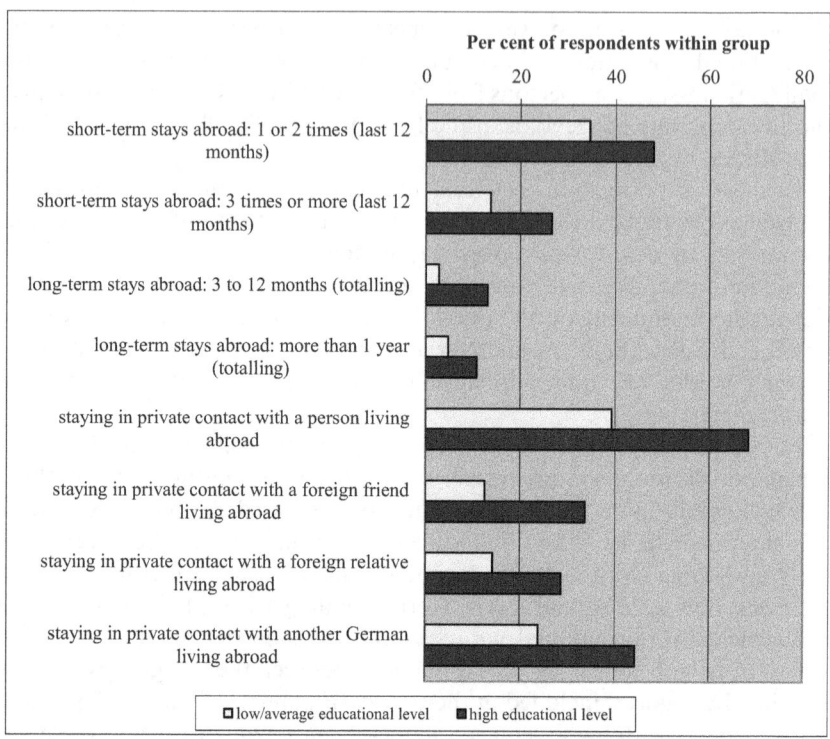

Figure 8.1   Educational attainment and transnational mobility

Note: The figure shows the percentage of 'yes' responses per educational level. The findings are weighted on the person level; the classification refers to the target school qualification of respondents still attending school in the survey period.

Source: Survey Transnationalisierung 2006 (weighted data).

currently recognised by the United Nations. Essentially, these 20 countries are the core states of the OECD, so that we could refer polemically to a 'first world transnationalisation'. However, there is also a close correlation between the education of the respondents and the spatial size of their networks. We have illustrated this for the two educational groups (high educational level vs. low/average educational level) in two world maps, which show the percentage of each contact country in the transnational contacts as a whole (Figure 8.2).

These maps illustrate that the international connectivity of the persons with lower and average educational levels is more strongly concentrated on a small number of countries, and there is less geographical spread in comparison to the higher educated group. With a 1 per cent threshold, South America, South Africa and even Asian countries such as China and Japan no longer appear on the social map of this group. In actual fact it appears that the contacts of the

The German Population and its Cross-Border Practices 173

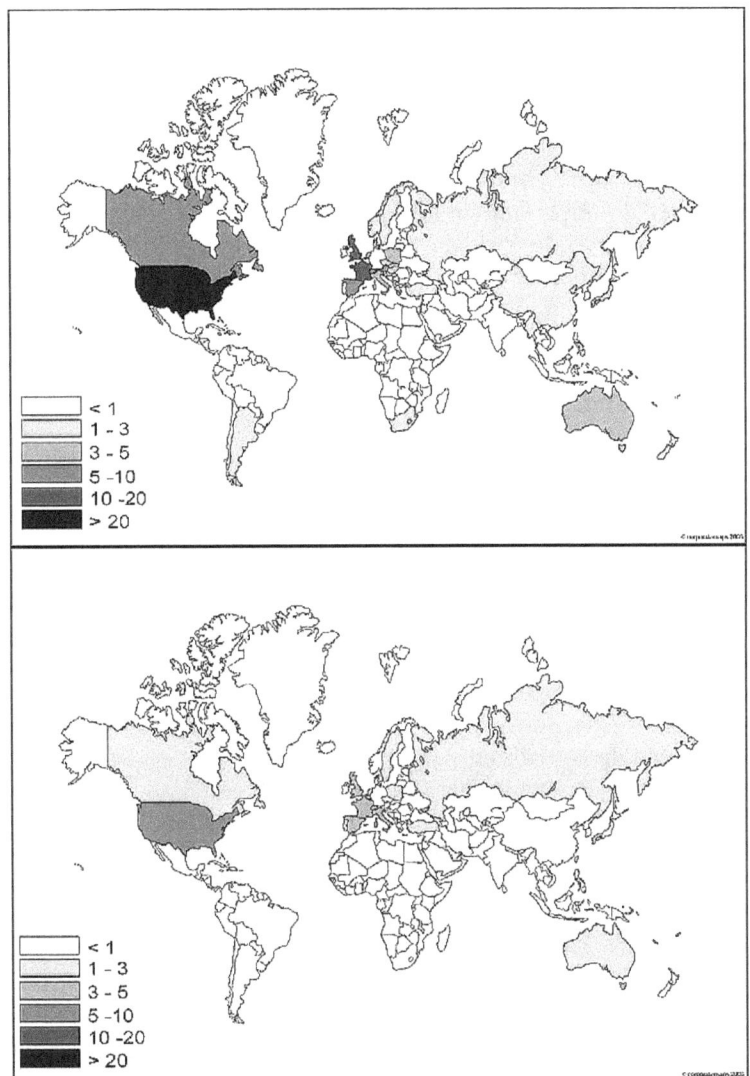

Figure 8.2   World maps by education: above – high educational level; below – low/average educational level

Note: The figure shows the percentage of those respondents who have regular contact with a person abroad and a transnational relationship with the country in question, weighted on the person level. To simplify the presentation, Greenland, Alaska, the northern islands of Russia and further sparsely populated regions are not marked.

Sources: *Survey Transnationalisierung 2006* (weighted data); base map: corporatemaps.

groups with low and average educational levels are not only sparser, but also cover a far smaller geographical area. There are far more blank spots on the map to which almost no social relationships exist than among the respondents with upper secondary level education.

We can also observe a discrepancy between the two educational groups with regard to transnational mobility, that is business and/or holiday trips made abroad as well as longer stays in foreign countries. Classifying the number of professional and private trips abroad undertaken by the interviewees in the past 12 months according to the respondents' respective schooling level, the persons with low and average educational levels are significantly less mobile. Whereas only a quarter of persons with upper secondary school-leaving certificates or higher education aptitude state that they did not go abroad at all in the past 12 months, this figure is 52 per cent among participants with a lower educational level. At the same time, the two groups differ in terms of the distance they cover in the course of their short-term trips made abroad (Figure 8.3). Again, it is the educationally rich whose journeys lead to the furthest places on average.

As regards longer stays abroad, the class-differences are even more drastic. Whereas only eight per cent of respondents with low and average educational levels state that they have lived abroad for at least three months in the past, this figure is three times as high among those with higher secondary school leaving certificates (24 per cent). In terms of the average distance between Germany and the respective target countries of stays abroad, we observe the same educationally structured pattern as we did in matters of shorter trips made beyond the German border (Figure 8.3).

With regard to the respondents' occupational status, we refer to the standard demography for telephone interviews (Statistisches Bundesamt 1999) as developed by the ZUMA, which is the Centre for Survey Research and Methodology (Mannheim, Germany).[6] We also take into consideration *previous* occupations if the respondents were not employed at the time of the interview. Complementary to the education-specific unequal transnationalisation, our data indicate that the higher occupational status groups are also far more broadly involved in transnationalisation than the lower-status groups. For example, only 32 per cent of unskilled workers state that they communicate regularly with a person abroad, whereas 68 per cent of executive and administrative-grade government employees and 65 per cent

---

6   With regard to the occupational status, we both consider the terms of employment (workers, employees, self-employed, civil service, farmers) as well as the position in the respective occupational group. Altogether, our analyses took into consideration nine different occupational status groups: workers (1. unskilled manual workers; 2. skilled manual workers/ master craftsmen), employees (3. low-skilled employees, 4. skilled employees, 5. managers), civil-service (6. lower/clerical grade, 7. administrative/ executive grade), self-employed (8. without employees, 9. with employees).

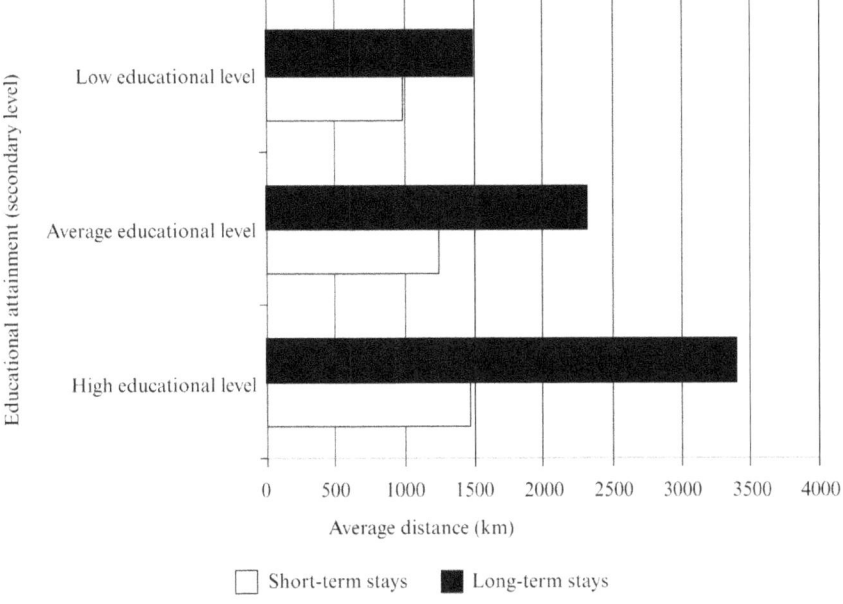

Figure 8.3   Average distance of transnational short- and long-term mobility (by educational level)

Note: Due to small case numbers, data for respondents without secondary school qualifications are not presented. The survey contains information about at most five target countries of short-term stays abroad (during the previous 12 months) as well as about at most four countries in which the given respondent has possibly been living during her or his life. In order to approximately compute the average spatial range of the respondents' stays abroad, we calculated the distance between Kassel (situated in the heart of Germany) and the respective capital of the country in question.

Source: Survey Transnationalisierung 2006 (weighted data).

of the self-employed with employees claim to do so. Again, the most marked differences are in non-family ties. Among workers, 8 per cent (unskilled workers) and 13 per cent (skilled workers/ master craftsmen) state that they have private contact with friends and acquaintances outside of Germany. In contrast, almost a quarter of managers and self-employed persons with employees have such private contact. Government employees in the executive and administrative grades have the highest transnational orientation in their networks of friends and acquaintances: more than one third (38 per cent) of respondents in this group state that they are acquainted or friendly with at least one non-German living outside of Germany. In line with the previously discussed hypotheses from the field of network research, we can thus observe that the reach of social networks increases with higher social status.

This pattern of unequal transnationalisation is repeated on the level of individual mobility. Be it the number of trips abroad or the geographical differentiation, the higher status groups prove clearly more transnational with regard to all indicators. Comparing the number of short-term stays abroad within one year (professional and/or private), more than of half the interviewed skilled workers and master craftsmen stated that they did *not* cross the German border at all, as did 70 per cent of unskilled workers. Managers, for example, are considerably more mobile (one to two trips abroad: 30 per cent, three or more: 30 per cent), as are executive and administrative-grade government employees (one to two trips abroad: 57 per cent; three or more: 31 per cent). The analysis of long-term stays abroad, that is stays of at least three months in a foreign country, provides similar findings: in this case, large parts of the groups of managers, executive and administrative-grade government employees and the self-employed with employees in particular can look back on experiences of living abroad.

Against the background of these descriptive findings, we carried out a multivariate OLS-regression analysis to provide information on how strongly involvement in transnational fields of interaction is influenced by socio-structural determinants. We constructed an index that measures the individuals' involvement in transnational social practices, containing the number of cross-border private relationships (contacts to non-German acquaintances and relatives living outside Germany and to Germans living abroad) and the respondents' short-term and long-term stays abroad. The construction of an index allows weighting its constituting factors unequally. We made use of that possibility, because we assume that life course passages spent abroad (i.e., all stays in foreign countries lasting more than three months) influence the individuals' lives more strongly than holiday trips, for example. So we decided to put stronger weight on the factor of long-term mobility. The rounded index covers a value range from 0 to 10, whereby '0' indicates that the respondent is entirely locally or nationally anchored in terms of his or her mobility behaviour and interpersonal relationships. Interviewees with an index value of '10', in contrast, are extremely highly involved in cross-border social networks and go abroad particularly often. We use this index as the *dependent variable* in our regression models (for a detailed description of the index, see Mau et al. 2008, 23–4).

With regard to the covariates used in our models, we measure education by years of schooling until graduation (value range: +8 to +13). In cases in which respondents were still attending school at the time of the interview, we have used the expected years at school. Concerning occupational status, we use the same set of categories as in our descriptive analysis. For capturing the relationship between the independent and the dependent variables adequately we have to control for a number of other sociodemographic factors. Since there is most likely a strong *East-West divide* which may interact with our determinants of interest, we also control for this: people in the former GDR (East Germany) had far more difficulties making contact with people from abroad than the citizens from West Germany. We measure the East-West classification with the question: 'Where did you live at the time of the German reunification?'[7] Furthermore, we follow Edmunds and Turner (2005, 572), who assume that we are currently witnessing the *formation of a new global generation* that 'both shares its information and ideas across borders and acts with global impact'. In this way, we assume that the younger the people are, the more strongly they are integrated into transnational spaces of interaction and mobility.

Also, we assume that we would find *gendered* patterns of transnational interaction and border-crossing mobility. Women are still more occupied with domestic responsibilities than men, regardless of whether they are economically active or not (Turner and Grieco 2000). This might reduce women's chances to make contacts to people from different contexts than household, neighbourhood and family (see Allan 1979, 1989). From research on mobility and transportation, we also know that women's trips are confined within a smaller geographical area than those of men (Jones et al. 1983). Hence, we expect men to be more strongly involved in transnational social practices than women.

Moreover, the global city debate (for example, Owen 1989; Sassen 1991, 2002) calls attention to the outstanding role of urban and metropolitan areas in the process of globalisation. In this way, large cities seem to be places 'where the work of globalisation gets done' (Sassen 2003, 13). We use community size (derived from the German BIK classification) as a proxy for the degree of urbanisation, assuming that people from dense populated parts of the country display comparatively high levels of involvement in transnational social practices.

A first bivariate regression with the transnationalisation index as a dependent variable and the respondents' educational level as an independent variable initially reveals a highly significant and positive correlation (Table 8.2, Model I). Just taking into consideration the respondents' school performance at this stage of our analysis, Model I is able to explain about 12 per cent of the variance. Thus, the assumption that educational level is a good predictor for the individual grade of transnationalisation is initially confirmed.

Another indication for more transnational involvement of the societal elites is to be found in the correlation between the grade of transnationalisation and the respective occupational status (Table 8.1, Model II). Respondents occupied in low-status positions are significantly less involved in cross-border activities and relations than those occupying higher-status positions. Compared to the

---

7   People who were living outside of Germany in 1990 were not included in the analysis, as we were unable to reconstruct whether these respondents were naturalised foreigners or Germans living abroad at the time.

Table 8.1   Determinants of involvement into transnational social practices (OLS regressions)

Variables	Model I		Model II		Model III		Model IV	
Years of schooling	.41	***					.33	***
Unskilled manual workers			-.44	**			-.43	*
Skilled person. Skilled orkers/ mastercraftspeople *(Ref.)*			*(Ref.)*				*(Ref.)*	
Low-skilled employees			.19				.15	
Skilled employees			.76	**			.46	***
Managers (executive employees)			1.33	***			.95	***
Civil service: lower/clerical grade			.69	*			.29	
Civil service: executive/ administrative grade			2.16	***			.92	***
Self-employed without employees			.74	***			.32	
Self-employed with employees			1.34	***			.90	***
Cohort 1915–1932					*(Ref.)*		*(Ref.)*	
Cohort 1933–1945					1.04	***	1.00	***
Cohort 1946–1967					.94	***	.78	***
Cohort 1968–1990					.96	***	.79	***
Gender: male *(ref.: female)*					.33	***	.29	***
Respondent lived in West Germany (FDR) in 1990 *(ref.: East Germany)*					.39	***	.48	***
Community size: <100,000 inhabitants					*(Ref.)*		*(Ref.)*	
100,000–499,999					.67	***	.34	***
≥500,000					.48	***	.41	***
Constant			1.36	***	1.95		-3.31	***
R² (adjusted)	.118		.095		.053		.198	

Notes:
Case number n=2273. The findings are weighted on the individual level. The dependent variable is the transnationalisation index, consisting of three components. The unstandardised regression coefficients are stated.
(Ref.) indicates the reference category of the variables in question.
*** sig. at p<0.001; **sig. at p<0.01; *significant at p<0.05.

Source: Survey Transnationalisierung 2006.

occupational group of skilled workers and master craftsmen, Model II shows a stronger integration in everyday cross-border contexts, at a high significance level, for managers, executive and administrative-grade government employees and the self-employed with employees. Unskilled workers, on the other hand, are significantly *less* transnationalised in comparison to higher-status occupational groups. Yet, the people's occupational status explains less variance (9.5 per cent) than their schooling.

To monitor whether the two central independent variables – education and occupational status group – do make a significant contribution to clarifying the variance with relation to transnationalisation, Model III estimates the sole influence of the sociodemographic control variants used. As we are mainly interested in involving all the variables described, we will not go into a discussion of the findings at this point.

Finally, Model IV (Table 8.1) shows that the observed effects of educational attainment and occupational status remain stable, strong and significant when the control variables are introduced. The respective educational level achieved thus structures the involvement in transnational interactions and activities to a high level. Further, we can see that transnationalisation is closely linked with occupational status. Even if the multivariate regression only reproduces the effect sizes towards the control group of skilled workers/master craftsmen, we can take the descriptive findings discussed above into account to conclude that higher-status working activities (regardless of the form of employment) obviously open up privileged access to the markets of opportunity both for making transnational contacts and being transnationally mobile.

With regard to the sociodemographic control variables, we can further observe a gender difference: men are more transnationally mobile and simultaneously more strongly involved in cross-border social relations than women. Even when controlling for whether the respondents were economically active or doing housework at the time of the interview, we still observe a clear gender divide with regard to the degree of involvement in transnational social practices. This result clearly strengthens the thesis of gendered patterns of mobility and making contacts. Additionally, there is still a gap between the respondents who were living in West Germany at the time of reunification and those who were living in East Germany at this time. Persons who were resident in the Federal Republic in 1990 are more strongly involved in transnational arenas. A further structuring element of integration in processes of micro-transnationalisation is the size of the respective community: the larger its population, the stronger the participation in transnational communication and the more intensive the experience of spending time abroad. Taking into account that Model IV explains about 20 per cent of variance, we may conclude that the used measures of social inequality make good predictors of individual involvement in transnational social practices.

As a whole, the findings support the hypothesis of unequal transnationalisation. By and large, the involvement in transnational practices, social relations and forms of dialogue is not a purely elite phenomenon, as the processes of transnationalisation have long since reached the general public. However, if we enquire not only into whether transnationalisation takes place but also into its extent, clear class differences come to light. Compared to the educationally poor and low-status groups, which tend to enter the transnational arena on a rather sporadic basis, the high-status groups place more emphasis on breaking down the national borders of their everyday lives: they have more contacts abroad (particularly elective ties), they cross the German border more frequently and more of them spend long periods living abroad.

Conclusions

The processes of globalisation and denationalisation are contributing to sustainable changes in the social order of the nation state. As we have illustrated, these processes are not restricted to economic links, global exchanges of information, new ecological risks and political activities beyond the bounds of the nation state, but are exerting more and more influence over individuals' lives. The population itself is also massively participating in cross-border interaction. From the perspective of 'transnationalisation from below', which takes account of both individual cross-border network ties and mobilities, this chapter examined whether we can actually assume a transnational mobilisation across all population strata, or whether there are major class differences. Our initial hypothesis was that the higher-status and better educated groups are more frequently and simultaneously more intensively active on a transnational basis, due to their functional involvement in global dialogue processes and due to their greater transnational competence. The findings show, firstly, that transnational experiences have now become very much part of everyday life and that large groups of the population cross national borders as a matter of course in their social relations and their own spatial mobility (see also Mau and Mewes 2007). However, we can also observe that this involvement is not equally distributed across the social structure. The groups with higher status and educational levels may actually be regarded as 'pioneers' of transnationalisation. Their social networks frequently stretch across national borders and they are significantly more active in terms of transnational mobility. The groups with lower occupational positions and lower educational levels, in contrast, are involved in transnational activities to a comparatively minor extent and are more fixed in their own nation state.

For the future development of the social structure, this may mean that the spatial horizons of the individual groups will become more differentiated and we will have larger sections of the population outside of the old national coordinates. This harbours the risk that the life horizons of differing groups, once united by the impermeable boundaries of the nation state, may drift apart. This is also the reason why some analyses from political sciences put forward the thesis that we will face a shift in the lines of conflict of national societies in the long term: new political conflicts may flare up due to the tension between an opening-up and internationalisation on the one hand and calls for re-closure with tendencies of re-nationalisation on the other (Kriesi and Grande 2004). This impact will presumably be cushioned to the extent that educationally poor and low-status groups become involved in movements of transnationalisation. There are a number of indications for such a development. However, they also show that there are very class-specific forms of transnational experience. When it comes to questions of the change in mentality and the change of political orientations connected with such a development, there is still an extensive need for research, to ultimately understand the challenge represented by transnational mobilisation for the nation state's integration capacity.

References

Adams, R.G. and Allan, G. (eds) (1998), *Placing Friendship in Context* (Cambridge: Cambridge University Press).
Albrow, M. (1998), 'Auf dem Weg zu einer globalen Gesellschaft?', in Beck, U. (ed.), *Perspektiven. der Weltgesellschaft* (Frankfurt/M.: Suhrkamp).
Allan, G. (1979), *A Sociology of Friendship and Kinship* (London: George Allen & Unwin).
Allan, G. (1989), *Friendship. Developing a Sociological Perspective*, (New York: Harvester Wheatsheaf).
Allan, G. (1998), 'Friendship and the Private Sphere', in Adams, R.G. and Allan, G. (eds), *Placing Friendship in Context* (Cambridge: Cambridge University Press).
Axhausen, K.W. (2007), 'Activity Spaces, Biographies, Social Networks and their Welfare Gains and Externalities. Some Hypotheses and Empirical Results', *Mobilities* 2:1, 15–36.
Axhausen, K.W. and Frei, A. (2007): 'Contacts in a Shrunken World'. Arbeitsbericht Verkehrs- und Raumplanung ', <http://www.ivt.ethz.ch/vpl/publications/reports>, accessed 24 April 2008.
Bauman, Z. (1998), *Globalisation. The Human Consequences* (New York: Columbia University Press).
Beaverstock, J.V. (2005), 'Transnational Elites in the City: British Highly-Skilled Inter-Company Transferees in New York City's Financial District', *Journal of Ethnic and Migration Studies* 31:2, 245–68.
Beck, U. and Lau, C. (eds) (2004), *Entgrenzung und Entscheidung* (Frankfurt/M.: Suhrkamp).
Beisheim, M., Dreher, S., Walter, G., Zangl, B. and Zürn, M. (1999), *Im Zeitalter der Globalisierung? Thesen und Daten zur gesellschaftlichen und politischen Denationalisierung* (Baden-Baden: Nomos).

Boissevain, J. (1974), *Friends of Friends. Networks, Manipulators and Coalitions* (Oxford: Basil Blackwell).

Brinkschröder, M. (1999), 'Klassenstruktur und Vergemeinschaftung im Postfordismus', in Rademacher, C., Schroer, M. and Weichens, P. (eds), *Spiel ohne Grenzen? Ambivalenzen der Globalisierung* (Opladen: Westdeutscher Verlag).

Carroll, W.K. and Fennema, M. (2002), 'Is There a Transnational Business Community?', *International Sociology* 17:3, 393–419.

Cass, N., Shove, E. and Urry, J. (2005), 'Social Exclusion, Mobility and Access', *Sociological Review* 53:3, 539–55.

Edmunds, J. and Turner, B.S. (2005), 'Global Generations: Social Change in the 20th Century', *British Journal of Sociology* 56:4, 559-77.

Favell, A. et al. (2006), 'The Human Face of Global Mobility: A Research Agenda', in Smith, M.P. and Favell, A. (eds), *The Human Face of Global Mobility. International Highly Skilled Migration in Europe, North America and the Asia-Pacific* (New Brunswick, NJ: Transaction).

Gerhards, J. and Rössel, J. (1999), 'Zur Transnationalisierung der Gesellschaft der Bundesrepublik. Entwicklungen, Ursachen und mögliche Folgen für die europäische Integration', *Zeitschrift für Soziologie* 28:5, 325–44.

Giddens, A. (1990), *The Consequences of Modernity* (Stanford, CA: Stanford University Press).

Greve, J. and Heintz, B. (2005), 'Die 'Entdeckung' der Weltgesellschaft: Entstehung und Grenzen der Weltgesellschaftstheorie', in Heintz, B. et al. (eds).

Guarnizo, L.E. and Smith, M.P. (1998),The Locations of Transnationalism', in Smith, M.P. and Guarnizo, L.E. (eds).

Hannerz, U. (1996), *Transnational Connections. Culture, People, Places* (London: Routledge).

Heintz, B., Münch, R. and Tyrell, H. (eds) (2005), *Weltgesellschaft. Theoretische Zugänge und empirische Problemlagen* (Stuttgart: Lucius & Lucius).

Heintz, P. (1976), 'Sozio-ökonomische und politische Indikatoren für die Beschreibung der Weltgesellschaft: Eine allgemeine Darstellung des Problems', in Hoffmann-Nowotny, H.J. (ed.), *Soziale Indikatoren. Internationale Beiträge zu einer neuen praxisorientierten Forschungsrichtung.* (Frauenfeld: Huber).

Heintz, P. (1982), *Die Weltgesellschaft im Spiegel von Ereignissen* (Diessenhofen: Rügger).

Hoffmann-Nowotny, H.J. (ed.) (1976), *Soziale Indikatoren. Internationale Beiträge zu einer neuen praxisorientierten Forschungsrichtung.* (Frauenfeld: Huber).

Jones, P.M., Dix, M.C., Clarke, M.I. and Heggie, I.G. (1983), *Understanding Travel Behaviour* (Aldershot: Gower).

Kaufmann, V. (2002), *Re-thinking Mobility. Contemporary Sociology* (Aldershot: Ashgate).

Kennedy, P. (2004), 'Making Global Society: Friendship Networks among Transnational Professionals in the Building Design Industry', *Global Networks* 4:2, 157–80.
Knight, R.V. and Gappert, G. (eds) (1989), *Cities in a Global Society. Urban Affairs Annual Review* (London: Sage).
Koehn, P.H. and Rosenau, J.N. (2002), 'Transnational Competence in an Emergent Epoch', *International Studies Perspectives* 3:2, 105–27.
Konrad, G. (1984), *Antipolitics* (New York: Jovanovic).
Kriesi, H. and Grande, E. (2004), 'Nationaler politischer Wandel in entgrenzten Räumen', in Beck, U. and Lau, C. (eds) (2004), *Entgrenzung und Entscheidung* (Frankfurt/M.: Suhrkamp).
Larsen, J., Urry, J. and Axhausen, K.W. (2006), *Mobilities, Networks, Geographies* (Aldershot: Ashgate).
Luhmann, N. (1997), *Die Gesellschaft der Gesellschaft* (Frankfurt/M.: Suhrkamp).
MacDonald, R., Shildrick, T., Webster, C. and Simpson, D. (2005), 'Growing Up in Poor Neighbourhoods. The Significance of Class and Space in the Extended Transitions of 'Socially Excluded' Young Adults', *Sociology* 39:5, 873–91.
Marsden, P.V. (1987), 'Core Discussion Networks of Americans', *American Sociological Review* 52:1, 122–31.
Mau, S. (2007), *Transnationale Vergesellschaftung. Die Entgrenzung sozialer Lebenswelten* (Frankfurt/M.: Campus).
Mau, S. and Mewes, J. (2007), 'Transnationale soziale Beziehungen. Eine Kartographie der deutschen Bevölkerung', *Soziale Welt* 58:2, 207–26.
Mau, S., Mewes, J. and Zimmerman, A. (2008), 'Cosmopolitan Attitudes through Transnational Social Practices?', *Global Networks* 8:1, 1–24.
Merton, R.K. (1995), 'Einflussmuster: Lokale und kosmopolitische Einflussreiche', in Merton, R.K. (ed.), *Soziologische Theorie und soziale Struktur* (New York: De Gruyter).
Merton, R.K. (ed.) (1995), *Soziologische Theorie und soziale Struktur* (New York: De Gruyter).
Nowicka, M. (2006), *Transnational Professionals and their Cosmopolitan Universes* (New York: Campus).
Ohnmacht, T., Frei, A. and Axhausen, K.W. (2008), 'Mobilitätsbiografie und Netzwerkgeografie: Wessen soziales Netzwerk ist räumlich dispers?', *Swiss Journal for Sociology* 31:1, 131–164.
Owen, W. (1989) 'Mobility and the Metropolis', in Knight, R.V. and Gappert, G. (eds), *Cities in a Global Society. Urban Affairs Annual Review* (London: Sage).
Pries, L. (2002), 'Transnationalisierung der sozialen Welt?', *Berliner Journal für Soziologie* 12:2, 263–73.
Rademacher, C., Schroer, M. and Weichens, P. (eds) (1999), *Spiel ohne Grenzen? Ambivalenzen der Globalisierung* (Opladen: Westdeutscher Verlag).

Sassen, S. (1991), *The Global City. New York, London, Tokyo* (Princeton, NJ: Princeton University Press).
Sassen, S. (ed.) (2002), *Global Networks, Linked Cities* (London: Routledge).
Sassen, S. (2003), 'Globalisation or Denationalisation?', *Review of International Political Economy* 10:1, 1–22.
Schmidt, G. and Trinczek, R. (eds) (1999), *Globalisierung. Ökonomische und soziale Herausforderungen am Ende des zwanzigsten Jahrhunderts* (Baden-Baden: Nomos).
Sklair, L. (1991), *Sociology of the World System* (London: Prentice Hall).
Sklair, L. (2001), *The Transnational Capitalist Class* (Oxford: Blackwell).
Smith, M.P. and Guarnizo, L.E. (eds) (1998), *Transnationalism from Below* (New Brunswick, NJ: Transaction).
Smith, M.P. and Favell, A. (eds) (2006), *The Human Face of Global Mobility. International Highly Skilled Migration in Europe, North America and the Asia-Pacific* (New Brunswick, NJ: Transaction).
Statistisches Bundesamt (1999), 'Demografische Standards', <http://www.gesis.org/Methodenberatung/Untersuchungsplanung/Standarddemografie/dem_standards/demsta99.pdf>, accessed 29 November 2007.
Stichweh, R. (1999), 'Globalisierung der Wissenschaft und die Region Europa', in Schmidt, G. and Trinczek, R. (eds), *Globalisierung. Ökonomische und soziale Herausforderungen am Ende des zwanzigsten Jahrhunderts* (Baden-Baden: Nomos).
Stichweh, R. (2000), *Die Weltgesellschaft. Soziologische Analysen* (Frankfurt/M.: Suhrkamp).
Swaan, A. de (1995), 'Die soziologische Untersuchung der transnationalen Gesellschaft', *Journal für Sozialforschung* 35:2, 107–20.
Szanton Blanc, C., Basch, L. and Glick Schiller, N. (1995), 'Transnationalism, Nation-States, and Culture', *Current Anthropology* 36:4, 683–6.
Turner, J. and Grieco, M. (2000), 'Gender and Time Poverty: The Neglected Social Policy Implications of Gendered Time, Transport and Travel', *Time Society* 9:1, 129–36.
Urry, J. (2004), 'Small Worlds and the New "Social Physics"', *Global Networks* 4:2, 109–30.
Urry, J. (2007), *Mobilities* (Cambridge: Polity).
Vobruba, G. (1995), 'Die soziale Dynamik von Wohlstandsgefällen. Prolegomena zur Transnationalisierung der Soziologie', *Soziale Welt* 46:3, 326–41.
Völker, B. and Flap, H. (2007), 'Community at the Workplace', in Lüdicke, J. and Diewald, M. (eds), *Soziale Netzwerke und soziale Ungleichheit. Zur Rolle von Sozialkapital in modernen Gesellschaften* (Wiesbaden: VS Verlag für Sozialwissenschaften), 113–34.
Watts, D.J. (1999), *Small Worlds* (Princeton, NJ: Princeton University Press).
Watts, D.J. (2003), *Six Degrees – The Science of a Connected Age* (New York: Norton).

Wellman, B., Carrington, P. and Hall, A. (1988), 'Networks as Personal Communities', in Wellman, B. and Berkowitz, S.D. (eds), *Social Structures: A Network Approach* (Cambridge: Cambridge University Press).

Wellman, B. and Berkowitz, S.D. (eds) (1988), *Social Structures: A Network Approach* (Cambridge: Cambridge University Press).

Willmott, P. (1987), *Friendship Networks and Social Support* (London: Policy Studies Institute).

Wittel, A. (2001), 'Towards a Network Sociality', *Theory, Culture and Society* 18:6, 51–76.

Zürn, M. (1998), *Regieren jenseits des Nationalstaats* (Frankfurt/M.: Suhrkamp).

This page has been left blank intentionally

Chapter 9
Residential Location, Mobility and the City: Mediating and Reproducing Social Inequity

Markus Hesse and Joachim Scheiner

Prologue

It was a natural disaster and its impact on human beings and society that unearthed evidently the socially inequitable distribution of mobility opportunities in modern societies. When the hurricane Katrina hit New Orleans and the Gulf Coast of Louisiana in the USA on 29 August 2005, it clearly demonstrated what dramatic and discriminatory consequences the limited possibilities for mobility can have in these days. It quickly became obvious that part of the (white) middle and upper classes were able literally to escape from the floods by car while African-American neighbourhoods without access to private motorised transport and without a functioning public transport system were practically abandoned to the flood waters. Consequently, people who lacked access to transport systems were coined 'the mobility poor' (Cresswell 2006, 259ff), indicating that the right or the ability to move is becoming increasingly important in the late-modern society. Cresswell notes that whereas about 85 per cent of the population, predominantly white and middle class, had already left New Orleans before Katrina struck, over 77,000 households comprising about 200,000 people remained in the city without any opportunity to get out of it (ibid.).

The uneven distribution of mobility chances was obvious before, but city officials were well aware of the fact that in the case of disaster many people would not be able to become evacuated or even to rescue themselves, respectively. So Katrina helped to illuminate a particular property of the modern city: that not only space (in terms of housing and location) is subject to social and economic inequity, but also the management of space-time relations, expressed by the opportunities of getting around and practicing circular mobility.

In the following, we aim at further exploring this particular role of mobility in (uneven) urban development. In so doing, the chapter ties up to a certain subject of classical urban sociology: the impact of urban form on mobility, the related spatial structure and social relations that are mutually intertwined. According to Burgess (1925, cit. in Cresswell 2006, 36), mobility was not only a central determinant of urban growth, yet also produced a certain disorder to city and society. 'It is the fact of locomotion ... that defines the very nature of society. But in order that there may be any permanences and progress in society the individuals who compose it must be located' (Park 1925, after Cresswell 2006, 37).

The two angles of this relationship, housing and mobility and their respective significance for social inequity will be further explored in this chapter. Whereas much emphasis has been placed in the more recent past on the attempt to integrate the physical planning of sites and circulation, we are now interested in the ways how social and spatial inequity is being produced and reproduced by the very particular role of mobility. In empirical terms we present multivariate models of residential location choice. Our findings emerged out of the StadtLeben research project that was carried out in the Cologne region between 2002 and 2005 (see Beckmann et al. 2006).

Social Inequity, Urban Development and Mobility

Questions of social inequity and the spatially inequitable distribution of affluence, quality of life, environmental burdens and life contingencies have been the focus of urban research for decades. Theoretical approaches and far-reaching empirical studies are available, especially in segregation research (cf. Friedrichs 1995, 79ff) and gentrification studies (cf. Blasius and Dangschat 1990). Differentiated descriptive and explanatory concepts of the socio-spatial classification of urban areas have been developed, showing the spatial distribution of the population by social and ethnic status. The concentration of socially deprived population groups in certain urban areas (e.g., inner-city neighbourhoods), whose dynamic change in the process of up- and down-grading and the concomitant conflicts is now one of the determining patterns of urban development and urban structure (cf. Heitmeyer et al. 1998).

While such studies long concentrated on housing stock, social infrastructure and the quality of the public space, etc., mobility and transport/traffic have only recently attracted attention. From the 1980s and 1990s, transport development plans were drawn up in Germany with both environmentally sound *and* socially balanced urban development in mind. They were influenced not least of all by the fact that socially deprived sections of the population are highly exposed to the adverse impacts of motorised transport, e.g., along busy main roads. The goals adopted, such as congestion reduction and the urban integration of traffic facilities related to this socially inequitable pattern of nuisance distribution.

With the advent of mobility research grounded in the social sciences, the perspective on these problems has broadened (cf. Bonss and Kesselring 2001). Mobility is seen not only as a means of physical motion but also an important prerequisite for participating in society. This does not only apply to issues such as life situation, lifestyle or social status, yet is also effective in very practical means, e.g. once somebody's driver's licence is being revoked as a consequence of administrative offence or criminal proceedings. Regarding the uneven distribution of income and related chances of societal participation, spatial mobility is now also discussed under the headings social exclusion and social inclusion. In the United Kingdom, a commission was set up by the Labour government in the late 1990s to address this subject (DETR 2000; see also Lyons 2003; Preston and Rajé 2007). Among the aspects of social exclusion that Lyons enumerates (2003, 340) are the decline of public facilities and services, the discrepancy between individual aspirations and the given possibilities, as well as the deprivation of individuals (although the dividing line between social exclusion and inclusion is blurred). In this context, the particular importance of mobility is that pedestrian access is often no longer available; accessibility depends on mobility facilities. This means that access to opportunities depends on the availability of transport. In the United States, too, these problems have long been recognised and much discussed, for instance with regard to the obstacles that non-motorised population groups have to overcome to participate in the labour market (cf. Bullard and Johnson 1997).

In an increasingly flexibilised, mobile society, the problems of social exclusion are compounded by inadequate access to mobility. In peripheral rural regions, where public transport falls below the threshold of economic viability, mobility is considerably restricted, at least for the non-motorised sections of the population (e.g., older people without a car or driving licence) (cf. Gray et al. 2006). This problem could be exacerbated in the near future by demographic change, once the loss of population makes infrastructure and services provision falling below minimum standards, so certain regions or particular parts of society are becoming excluded from the use of amenities.

In urban areas, social inequity is evident in terms of social deprivation that may occur both caused by the lack of access to mobility and as a consequence of mobility-related degradation of living conditions (e.g. caused by air pollution, noise emissions etc.). Many cities in Europe still have highly stressed neighbourhoods and traffic corridors, which also have a high concentration of population groups with a low rate of motorisation or who are badly served by public transport. In such areas, the two categories of problem overlap: mobility (especially motorised transport) becomes a risk, contributing to the deterioration of living conditions; and the lack of mobility facilities prevents people from participating in society, limits access to education, the labour market, etc. Poverty and deprivation structures are thus mutually reinforced, also from a socio-spatial point of view.

This argument can be extended to discussing the delicate relationship between cities and suburbs: according to a classical understanding of suburbanisation, the outmigration to the suburbs represents a particular search for better living conditions and social homogeneity, which at the same time reinforces the degradation of living conditions for other people in other parts of the conurbation, notably the inner city. Since the locational change from city to suburbs and the emergence of more regionally oriented activity patterns are inevitably connected to the question of mobility (mostly its motorised version), the link between social inequity and mobility becomes even more evident. However, as suburbs, inner cities and the many different places in between are changing, so does the appearance and perception of social inequity. Some of the core urban areas are subject to renewal and upgrading, whereas others remain stagnating or continue to decline. Also suburbs are undergoing certain life cycles, according to age, life situation and preferences of their inhabitants. This particular change is already visible in the first and second generation of single-family suburbs e.g. in many Western European suburban areas, also in North America. In the case these suburbs are located in remote areas with low density and poor supply of infrastructure and services, demographic change, dropping real estate prices and rising energy costs may contribute to new forms of suburban social inequity (cf. Hesse and Scheiner 2007). This indicates that the issue of social inequity related to mobility turns out to be highly relevant in urban regards.

Evidence from the Cologne Region

Data and Research Outline

This section presents an empirical study of the interrelations sketched above. Spatial segregation as well as residential location behaviour of population segments defined by life situation and lifestyle are studied in binary logit regressions. Before the results are presented, the methodology is described subsequently.

The data used were collected in a household survey within the scope of the project StadtLeben.[1] The survey was undertaken in ten study areas in the region of Cologne in 2002 and 2003 (see Figure 9.1): 2,691 inhabitants took part in extensive face-to-face interviews about their travel behaviour, housing mobility, life situation, lifestyle, location preferences and residential satisfaction. The response rate was 27 per cent of those asked.

The region of Cologne is a polycentric agglomeration with the clearly dominating centre of Cologne. The demographic trend is slightly positive

---

1 'StadtLeben' – Integrated approach to lifestyles, residential milieux, space and time for a sustainable concept of mobility and cities' (2001–2005). Project partners: RWTH Aachen, Institute for Urban and Transport Planning (coordination); FU Berlin, Institute of Geographical Sciences, Urban Studies Unit; Ruhr-University of Bochum, Department of Cognition and Environmental Psychology; University of Dortmund, Department of Transport Planning (see http://www.isb.rwth-aachen.de/stadtleben/).

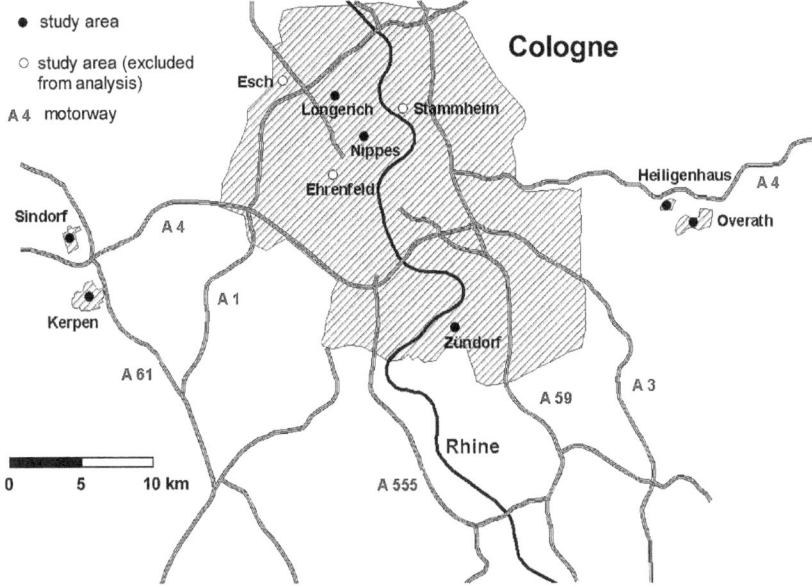

Figure 9.1   Location of the study areas in the region of Cologne

Source: Author's concept of project group StadtLeben.

and the housing market is largely supply dominated. The opportunities for different population groups as defined by lifestyle or life situation to realise a specific location choice that meets their needs and wishes are thus limited.

The study areas represent five area types, each type is represented by two areas: high-density inner-city quarters of the nineteenth century ('Wilhelminian style:' Ehrenfeld, Nippes); medium-density neighbourhoods dating from the 1960s ('modern functionalism') with flats in three- or four-storey row houses (Stammheim, Longerich); former villages located at the periphery of Cologne which since the 1950s have experienced ongoing expansion with single-family row houses or (semi-)detached single occupancy houses (Esch, Zündorf); small town centres in the suburban periphery of Cologne (Kerpen-Stadt, Overath-Stadt); and suburban neighbourhoods with detached single occupancy houses (Kerpen-Sindorf, Overath-Heiligenhaus). The four suburban neighbourhoods are all about 30 km away from Cologne.

As each of the two areas belonging to one type is clearly different, the areas are very varied with regard to spatial location, transport infrastructure, central place facilities and socio-demographic structure. Nonetheless it has to be noted that spatially or socially 'extreme' areas were not purposely targeted. There are no obvious high income areas and only one distinct low income area (Stammheim). In any case Stammheim along with Ehrenfeld and Esch are excluded from the analysis because the location preferences of the inhabitants of these areas could not be investigated due to reasons of project flow. The analysis is therefore based on the seven remaining study areas only. Furthermore, only adults aged 18 or older are included in the analysis. The resulting net samples vary between n = 1,572 and n = 524, depending on the model.

Variables

Residential location decisions are based on complex, multi-step decision-making processes in households, which include a primary decision on whether to move or not and a subsequent decision on the location to be chosen (Kim et al. 2005). In this section we are focusing on residential location choice, while we exclude the decision to migrate.

Location decisions are based on a large number of external conditions on the one hand and individual or household-related needs on the other hand. Our models focus on demand-related determinants of location choice: on individual resources and constraints, preferences and lifestyles. Supply-related aspects, such as the housing market, are beyond the scope of our analyses. The reason for this limitation is that our data were collected in a limited number of neighbourhoods. As a consequence, they do not allow for a thorough modelling of the decision that is made between all potential locations provided by the housing market in the region of Cologne. This means that we are focusing on the socio-spatial rather than the economical background of location choice.

In 'basic models' we use life situation and lifestyle as explanatory variables of location choice. Life situation variables include age, gender, household type, education level, household income per capita (with children counting as 0.8 persons), employment, job position (leading position yes/no) and nationality. All life situation indicators except for income are transformed into binary variables to achieve an appropriate scale level. In the case of age this allows the identification of non-linear associations with location choice. For instance, young adults tend to prefer central, inner-city locations, while middle age individuals tend to prefer suburban locations and the location preferences of retirees are less clear.

Due to expected interdependencies between various life situation indicators, interaction terms between age, education, income and employment have been included. As this results in rather large tables, the models presented in this chapter focus on main effects without interaction terms. The results are similar to the models with interaction terms that are available elsewhere (Scheiner 2005b).

Lifestyles are presented in the data by 34 Likert-type items representing three domains: leisure preferences, values and life aims, aesthetic taste. The items were reduced to eight lifestyle dimensions constructed as mean scales from the items. The construction of the scales was preceded by extensive factor analyses (Schweer and Hunecke 2005). As opposed to using discrete lifestyle types, this permits conclusions on which lifestyle attributes are most relevant for location choice. Additionally, the frequency of face-to-face contact to kin, neighbours, friends and colleagues was included, representing the social dimension of lifestyle. We expect high frequency of kin and neighbouring contact to represent a tendency towards 'nesting', which should be associated with suburban location choices. Vice versa, high contact frequency with friends and colleagues should be associated with a tendency towards urban locations. The dimensions and examples for items are shown in Table 9.1.

Table 9.1   Lifestyle dimensions and items

Domain	Dimension	Indicators/items (examples)
Leisure preferences	Out-of-home leisure	Frequency of going to cinema/theatre/concert Frequency of attending courses and education
	Domestic leisure	Frequency of playing with children Frequency of engaging with my family
Values	Self-realisation	Importance of leading an exciting life Importance of achieving a leading job position
	Traditional values	Importance of parsimony Importance of safety and security
Cultural schemes	Trivial	Interest in quiz programmes Interest in sentimental novels
	Suspense	Interest in action movies Interest in horror movies
	High culture (reading)	Interest in classical literature Interest in poems
	High culture (TV)	Interest in political features Interest in documentaries
Social networking	Kin networking	Frequency of face-to-face contact with kin
	Neighbour networking	Frequency of face-to-face contact with neighbours
	Friends and colleagues networking	Frequency of face-to-face contact with friends and colleagues

The 'basic models' are complemented by 'extended models' which include all the variables described above plus location preferences and the availability of transport options. Strong preferences for a quiet, green neighbourhood or the availability of a car, for instance, may be premises for suburbanisation (Scheiner 2006). By including preferences and travel options in the models, the question can be studied as to whether the relevance of lifestyles and life situations for location choice decreases as soon as other determinants are considered. This is a recent line of research in the debate on the role of residential self-selection for travel behaviour (Handy et al. 2005; Scheiner 2006).

Location preferences were operationalised using subjective importance ratings of neighbourhood and location attributes. Information was gathered as part of the survey by asking 'How important are the following features of the neighbourhood for your personal decision in favour of a certain place of residence?' The attributes were then listed, for instance 'accessibility of the city centre' or 'access to public transport'. The five-point Likert-type answer scales ranged from 'not important' to 'very important' and were constructed so that they came as close to an interval scale as possible (see Rohrmann 1978). The items were reduced to seven dimensions constructed as mean scales from the items (Table 9.2, see for details Scheiner 2006b). Again, the construction of the scales was preceded by extensive factor analyses.

Table 9.2   Location preference dimensions and items

Dimension	Indicators/items (examples)
Access to centre	Access to the city centre, access to the workplace
Proximity to shopping	Proximity to shopping facilities, proximity to services
Opportunities for children	Playgrounds for children, leisure opportunities for young people
Social neighbourhood	Neighbourhood, security from crime and vandalism
Residence	Size of the house/flat, quality of the building
Access to motorway	Access to motorway
Parking	Availability of parking spaces or garages

Car availability and the availability of a season ticket for public transport are measured by two binary variables (car in the household yes/no and season ticket yes/no).

The models described so far reflect social differences – the state of segregation – between an area type and all other study areas. Thus, we call these models 'state models'. These models do not necessarily reflect deliberate decisions for a certain residential location. They rather distinguish individuals living in a certain area from those who live elsewhere.

Additionally we estimate models with the same structure, including only those respondents who moved into an area not more than five years before the survey. The models study whether this move led to a certain neighbourhood type or not. This should more accurately reflect explicit decisions in favour of a certain area type. As the characteristics of the in-movers are likely to reflect ongoing changes in the population structure and therefore, socio-spatial transformations of the neighbourhoods studied, we call these models 'process models'. The period of five years was chosen as a compromise between the attempt to capture recent processes (which argues for a period as short as possible) and the limited sample size (which argues for a longer period). The total structure of the analyses undertaken is shown in Table 9.3.

Table 9.3   Structure of analysis: types of models

	State models: total sample	Process models: in-movers
Basic models: lifestyle, life situation	Basic state models	Basic process models
Extended models: lifestyle, life situation, transport options, location preferences	Extended state models	Extended process models

The dependent variables used are binary variables describing whether a respondent lives in a certain neighbourhood type or not. The neighbourhood types studied here are inner-city neighbourhoods and suburban neighbourhoods.

One has to keep in mind that due to the selection of only a limited number of neighbourhoods, the 'process models' distinguish whether a move led to a certain neighbourhood type or to another type among all areas studied. The models therefore do not compare one type with all other types existing, but with all other types studied. However, we believe that the selection of study areas represents a large range of neighbourhoods in a typical West-German agglomeration that is broad enough to allow thorough conclusions.

Methodology of Analysis

The above described variables were used in binary logistic regression models. These are based on a transformation of the binary endogenous variable. What is to be estimated is not the probability of a certain event (e.g. the choice of an inner-city neighbourhood), but the logarithmic ratio between this probability and the probability that the event does not happen ('Log-Odds') (Long 1997). This can be expressed as follows:

$$L(y=1) = \ln(P_1/(1-P_1)) = \alpha + \Sigma \beta_k x_k$$
$$\text{Odds ratio: } P_1/(1-P_1)$$
$$\text{Log-Odds: } \ln(P_1/(1-P_1))$$
$$\text{Logit coefficients: } \alpha, \beta_k$$
$$\text{exogenous variables: } x_k$$
$$\text{probability of the event: } P_1$$

Based on the estimated coefficients one can calculate the probability of the event for each respondent:

$$P_1 = \frac{\exp(\alpha+\Sigma\beta_k x_k)}{1 + \exp(\alpha+\Sigma\beta_k x_k)}$$

As we are mainly interested in the direction and strength of the associations between our exogenous variables and the respective endogenous variable, the result table gives effect coefficients $\exp(\beta_k)$. These quantify the marginal change in the odds ratio if an exogenous variable changes by one unit. Values larger than 1 increase the odds ratio, values smaller than 1 decrease it. The α-coefficients (intercepts) are less important in the context of this chapter. All analyses are done with unweighted data to avoid distortions of significance.

Results I: 'Basic Models' – The Role of Life Situation and Lifestyle

State models – The residential population of inner-city neighbourhoods distinctly differs from other areas mainly with respect to household type, gender, education level, ethnicity and lifestyle. Respondents with a university degree are more likely to live in inner-city neighbourhoods than others and the same is true for immigrants and women. All household types except for flat-sharing communities are less likely to live in an inner-city neighbourhood than single households.

Respondents aged 50 or older live in inner-city neighbourhoods less frequently than the reference category of those aged 18 to 29 does. Taking statistically insignificant effects into account makes clear that the inner-city is an area for young adults aged 18 to 39. There are distinct effects of lifestyle and they account for roughly a quarter of the explanatory value of the model, measured with the sum of the standardised effects (absolute deviation of effect coefficients from 1).

Individuals with a strong inclination to self-realisation, with high culture interests, or with frequent contact to friends and colleagues tend to live in the inner-city more often than others. By contrast, those with strong domestic leisure preferences, with a traditional value orientation or with frequent contact to their kin tend *not* to live there. This underlines the attractiveness of the dense, mixed-use 'Wilhelminian' areas for hedonistic, individualistic lifestyles with autonomously chosen, rather than pre-determined networking patterns.

Subsequently we turn our attention from the inner-city to suburbia. Household type has a pronounced effect on living in suburbia. Compared to single households, households with children, no matter whether complete families or single parents, are more likely to live in suburbia. Thus, single households turn out to appear as an 'urban' household type, as one might expect. Women are less likely to live in suburbia than men do, everything else being equal and the same is true for individuals who own a university degree, as compared to those who do not.

With respect to lifestyle, individuals with domestic leisure preferences and those with a strong inclination towards trivial culture are more likely to live in suburbia than others are, while individuals with frequent contact to friends and colleagues and those with a strong tendency towards self-realisation are less likely to do so.

Process models – The fit values of the state models improve considerably (R^2 inner-city from 0.273 to 0.415, suburbia from 0.135 to 0.283) once we limit our analysis to in-movers, as described above. According to this, location decisions are more differentiated in terms of demographic, socioeconomic and lifestyle structures than one might expect when looking at the total population in a given area.

Moving to the inner-city is strongly linked to young people, particularly singles less than 40 years of age. When statistically insignificant effects are taken into account one may even say, the inclination to move into an inner-city neighbourhood steadily decreases with age. It should be noted that this is not the result of an age-related general decrease in mobility, but reflects location decisions in favour of the inner-city as compared to other areas. What is more, moving into the inner-city is more likely for households without children, women, academics as compared to others, respectively.

With respect to lifestyle, high culture orientation and social networking patterns that are oriented towards friends and colleagues rather than kin, enhance the inclination to move to the inner city. By contrast, traditional values, domestic leisure preferences and a distinct orientation towards the trivial culture scheme decrease the likelihood of choosing an inner-city area. By contrast, moving to suburbia is significantly more often characterised by households with children (families, single parent households). Owning a university degree is negatively associated with moving to a suburban neighbourhood. Men tend to suburbanise more than women do. Age effects turn out statistically insignificant.

The directions of the lifestyle effects are essentially contrary to the inner-city model. Individuals with domestic leisure orientation and an inclination towards trivial culture are more inclined to suburbanise than others are, while individuals with a strong tendency towards self-realisation and friendship/colleagues-oriented networking patterns are less inclined to do so. One notable exception from the overall pattern is the positive association between out-of-home leisure preferences and suburban location choice. This suggests that moving to suburbia may not be quite as strongly linked to 'nesting' as it seems to be the case in some stereotypes.

Comparing effect strengths between the state models and the process models shows generally, although not in every case, stronger effects (i.e. larger deviation of the coefficients from 1) in the process models. This is reflected in better fit values, i.e. a higher proportion of variance explanation. Together with the general observation that the direction of the effects is consistent between each pair of state and process model, this suggests, firstly, that the in-movers resemble the old-established population of a certain area in terms of life situation and lifestyle. Secondly, the respective socio-spatial inequalities get even more distinct by incoming groups: succession of distinct, 'typical' population groups into the inner-city as well as into suburbia is recently on its way. In other words, residential mobility does not only reproduce, but even reinforce socio-spatial inequality.

Results II: 'Extended Models' – The Role of Location Preferences and Transport Options

Subsequently we turn our attention to the extended models that consider transport options and location preferences as additional variables. Possibly individuals choosing a particular place of residence or living at a particular place can be described more precisely this way, even we can not make conclusions on causality from cross-sectional data. This means that we cannot conclude on whether the availability of a vehicle or a public transport season ticket determines location choice or vice versa. The same is true for the causal flow between location preferences and location decisions.

The additional variables considerably improve the model fit values. The same is true again for limiting the analysis to in-movers in the process models as compared to the state models. The following interpretation focuses on the process models. Table 9.4 shows the state model results for comparison.

The direction of the effects generally confirms the results discussed above. Only in few insignificant cases the direction tends to change. To sum up the main results briefly: Moving to an inner-city neighbourhood is more likely for women than for men and more likely for academics. Models with interaction effects show that the latter is particularly true for academics aged 30 to 39 years (Scheiner 2005b). The lifestyle effects confirm the results presented in the former section as well. One additional lifestyle effect turns out to be significant for the first time: Individuals with a strong inclination towards suspense and thrill are less likely to move to the inner-city. This does probably not mean that the inner-city of Cologne is less thrilling than, say, a small town at the periphery. Rather, it reflects that the suspense scheme is not so much an aesthetic scheme of a highly skilled, culturally interested urban academic milieu than of the milieu consisting of young, fun-oriented people (mainly men, to be more specific) with medium education level.

What is more, the model shows the importance of subjective location preferences and transport options for residential location choice. Individuals who live in motorised households or/and who assign high relevance to access to a motorway are less inclined than others to move to the inner city. The opposite is true for those showing a high preference for proximity to shopping facilities in the neighbourhood.

For suburban areas, the effects of transport mode availability are vice versa. Individuals who live in motorised households and/or do not own a season ticket are more likely to suburbanise than others. The same is true for those who assign little importance to access to the centre and for individuals for whom access to a motorway is highly important. These accessibility preference

Table 9.4 Models of residential location choice – 'basic models' and 'extended models' vs 'state models' and 'process models'

	Basic models				Extended models			
	Inner city		Suburbia		Inner city		Suburbia	
	State model*	Process model*	State model*	Process model*	State model*	Process model*	State model*	Process model*
Age (ref.: 18–29)								
30–39	1.43	1.16	1.05	0.93	1.81	1.81	0.63	**0.52**
40–49	1.01	0.62	1.00	0.91	1.29	0.88	0.63	0.55
50–64	**0.50**	**0.32**	1.13	1.14	0.66	0.35	0.66	0.60
65+	0.57	0.19	0.80	0.46	**0.35**	**0.07**	0.57	0.48
Gender (female=1)	**2.04**	**2.19**	**0.69**	**0.57**	**1.94**	**2.33**	0.78	0.62
Household type (ref.: single)								
Family	**0.34**	**0.38**	**1.67**	**2.44**	**0.50**	0.40	1.26	**2.14**
Single parent	**0.30**	**0.15**	**2.40**	**3.71**	**0.37**	0.23	**2.22**	3.00
Couple w/o children	**0.69**	**0.73**	1.29	1.38	0.94	0.88	1.19	1.09
Flat-sharing	0.98	1.28	0.51	0.28	0.78	1.00	0.58	0.26
Education level (ref.: primary school or no degree)								
Secondary school I or II	0.78	1.17	0.95	0.66	0.96	1.64	0.76	0.56
University degree	**1.70**	**3.93**	**0.57**	**0.35**	**1.91**	**5.82**	**0.52**	**0.34**
Per capita household income	1.00	0.78	1.05	1.14	1.00	0.95	0.92	0.93
Employed (yes=1)	1.00	0.76	0.86	0.84	1.09	1.11	0.99	0.86
Nationality (non German=1)	**2.04**	**2.01**	**0.70**	**0.67**	1.90	1.74	0.78	0.89
Lifestyle								
Out-of-home leisure preference	1.07	0.83	1.09	**1.34**	1.12	0.84	1.04	**1.42**
Domestic leisure preference	**0.62**	**0.50**	**1.42**	**1.60**	**0.67**	**0.64**	**1.38**	**1.44**
Traditional values	**0.71**	**0.77**	1.07	0.97	**0.72**	**0.73**	1.08	0.99
Self-realisation	**1.24**	**1.26**	**0.85**	**0.78**	1.18	1.27	0.90	0.81
Trivial scheme	0.87	**0.69**	1.24	**1.67**	0.92	0.88	**1.24**	**1.70**
Suspense scheme	1.01	0.87	1.00	0.99	0.96	**0.70**	0.99	1.05
High culture scheme (TV)	1.13	**1.31**	1.03	0.94	**1.22**	1.35	0.96	0.99

Table 9.4 cont'd

	Basic models				Extended models			
	Inner city		Suburbia		Inner city		Suburbia	
	State model*	Process model*	State model*	Process model*	State model*	Process model*	State model*	Process model*
High culture scheme (reading)	**1.25**	1.15	**0.89**	0.92	**1.25**	1.14	0.95	0.99
Social network kin	**0.89**	**0.83**	0.98	1.06	**0.86**	**0.81**	0.98	1.03
Social network neighbours	0.96	0.94	0.96	1.04	0.97	0.93	0.97	1.01
Social network friends+colleagues	**1.11**	**1.24**	**0.93**	**0.91**	**1.08**	**1.17**	0.94	0.94
Car in household					**0.50**	**0.41**	**1.82**	1.50
Public transport season ticket					1.30	1.58	**0.44**	**0.38**
Location preferences								
Access to centre					1.24	1.27	**0.58**	**0.47**
Proximity to shopping					**1.71**	**2.04**	1.01	1.22
Social neighbourhood					0.78	0.81	0.83	0.67
Opportunities for children					1.02	0.92	1.06	1.12
Residence					**0.78**	0.98	**1.30**	1.19
Access to motorway					**0.64**	**0.53**	**1.32**	**1.44**
Parking					1.01	0.93	0.90	0.84
Constant	0.38	0.21	1.70	2.02	0.54	0.13	**6.63**	**31.74**
R² (Nagelkerke)	0.27	0.42	0.14	0.28	0.37	0.53	0.25	0.40
No.	1,572	601	1,572	601	1,336	524	1,336	524

Notes:

The table shows effect coefficients of logit models. Significant coefficients ($\alpha=0.05$) are bold. Values larger than 1 increase the odds ratio and represent a positive impact of the exogenous on the endogenous variable, whereas values smaller than 1 represent a negative impact.
* State models: total sample. Process models: Limited to in-movers into one of the study areas Grey: excluded from analysis.

Source: Author's analyses. Data: StadtLeben household survey.

effects seem obvious, but it is less obvious that they appear significant at all, given that a large range of social and lifestyle indicators are controlled for.

Summary of Empirical Findings

The results of the empirical study undertaken in the region of Cologne reveal considerable social inequities in the spatial distribution of an urban population and in location choices made in recent migration flows in an urban context. These inequities are to a large part based on structural attributes of the population, i.e. their life situations. Another significant part is shaped by lifestyles, which include leisure preferences, value orientations, aesthetic schemes and social networking patterns. The life situation as well as the lifestyle element are in line with segregation research, but two observations are still worth mentioning. First, we found generally consistent effects for each pair of state and process model for a given area type. This suggests that the existing socio-spatial distribution of a population is reproduced by residential mobility, i.e. by incoming new residents who resemble the old-established population in terms of their social structure as well as their lifestyles. Secondly, the effects found are often stronger for the process model than for the respective state model. This means that residential mobility does not only reproduce, but even reinforce socio-spatial inequality.

Beyond life situation and lifestyle, subjective preferences towards housing, the neighbourhood, travelling and access to facilities and transport infrastructure play a significant role for location choice. What is more, there are strong associations to the availability of transport options, be it a car or a season ticket for public transport. This is rarely recognised in the literature on segregation and residential choice. The car has become an autonomous structural attribute of households and individuals, which determines the options for location choices to a certain extent. The finding that ownership of a season ticket has a significant impact on location choice similar to the car – although in the opposite direction – leads us to the interpretation that the car effect need not be an outcome of the material artefact 'vehicle'. Rather, it could be a realisation of an attitude, a 'materialised preference:' car oriented individuals tend to live at peripheral locations, public transport oriented individuals tend to live at central locations.

Before concluding, two shortcomings of the data shall be highlighted. First, our analyses are based on individual data that do not adequately reflect interdependencies within households. For instance, location preferences may vary between household members and the location decisions actually made could be based on the preference of a household member who was not asked. As a consequence, the associations found between location preferences and location behaviour could be closer in reality than in our data, as reality is only partly reflected in our data. The same might be true for lifestyle.

Secondly, we did not take spatial differences in housing supply into account. Generally, the housing market in the Cologne region is largely supply dominated. The opportunities for different population groups as defined by lifestyle or life situation to realise a specific location choice that meets their needs and wishes are thus limited. However, we can not say much about intraregional differences. The income effects in our analyses are not quite as dominant as one might expect. Thus, cautiously speaking, the intraregional spatial differentiation of housing costs do not seem to play the most prominent role here.

A comparison between the extended models and the basic models suggests a remarkable reduction in the impact of life situation. While in the basic models life situation accounts for about 76 per cent of the explanatory value of all significant effects taken together (slightly varying between the models), this share declines to only 54 per cent on average in the extended models.

Although this is undoubtedly still a large proportion, the reduction suggests that socioeconomic and demographic inequalities in location choice are partly due to different location preferences of the respective population groups, rather than structural inequalities per se. The declining relevance of structural attributes as soon as preferences are accounted for has been shown before in the context of residential self-selection and travel behaviour (Kitamura et al. 1997, Handy et al. 2005).

What is more, the remarkable impact of the availability of transport options may be surprising. The availability of transport options accounts for 11 per cent of the explanatory value of all significant variables in the extended models on average. Eighteen per cent is due to location preferences, 17 per cent due to lifestyle and 54 per cent due to life situation. Beyond the well-established empirical knowledge on the importance of social inequity for location behaviour in terms of life situations and lifestyle differences, transport mode orientations or 'mobility styles' (Lanzendorf 2002) seem to be an important factor as well. However, it has to be highlighted at this point that the availability of transport options must not exclusively be interpreted as a determinant of location choice. Rather it appears to be a determinant as well as an outcome of location choice. People make their location decisions on the basis of transport options or travel mode orientations, but at the same time these orientations tend to change after relocations according to the conditions set by the new environment (Krizek 2003; Handy et al. 2005; Scheiner 2005a).

Conclusions, Policy Recommendations and Outlook

The findings of the StadtLeben project presented here confirm the important role that social conditions, particularly household configuration, life situation and also lifestyles play for explaining locational behaviour in urban regions. As a result, this extended view on certain facets of social inequity – by adding the lifestyle issue to more classical indicators of inequity, such as income and social status – further explores the diversity of late-modern urban development.

How can these findings be assessed in terms of policy? Is social inequity, if also evident in terms of leisure practice, normative orientation or social networks, good or bad for the contemporary city? Is it ultimately associated with urban life and thus necessarily to be taken into account, or should it become subject of programmes and policies to be managed, in favour of the recent *leitbild* of the 'integrated city?' The fair distribution of both quality of life and access to mobility might become an undisputed aim of any urban policy that claims to be sustainable and integrated. Regarding ambitious policy programmes and declarations for urban development – such as the Leipzig Charter on Sustainable European Cities, declared by the European Council of Ministers responsible for Urban Development in May 2007 – there is a strong momentum both towards social and environmental sustainability, thus supporting such a point of view: 'Cities are faced with major challenges, especially in connection with the change in economic and social structures and globalisation. Specific problems, among others, are especially high unemployment and social exclusion. Within one city, considerable differences may exist in terms of economic and social opportunities in the individual city areas, but also in terms of the varying quality of the environment. In addition, the social distinctions and the differences in economic development often continue to increase which contributes to destabilisation in cities. A policy of social integration which contributes to reducing inequalities and preventing social exclusion will be the best guarantee for maintaining security in our cities'. (European Council of Ministers responsible for Urban Development 2007, 5)

In this context, one of the essential conclusions and policy recommendations that may be derived from our empirical findings applies to the strong connections between i) social issues and mobility in general and ii) location choice and circular mobility in particular. These two fields of interrelation should be made subject to a coherent strategy in urban development policies, e.g. by sensitising transport planners and providers for the enormous significance of their businesses, also by linking land use and urban redevelopment planning with mobility and transport organisation and infrastructure.

How can this claim be enforced in more practical terms? Given the current extent of urban development dynamics in many of the European towns and cities, there are several urban 'areas' and related fields of action that generate a certain demand for strategy building:

- First, inner-city areas are subject to urban renewal and regeneration in a way that already provoked the assumption of an 'urban renaissance' that is apparently underway in major metropolises. Those areas might witness accelerated processes of segregation and gentrification and a related exchange of population. Given the life-style attributes of the particular social milieu of the gentrifiers, it is easy to predict that a rising

- Second, urban social inequities in general and the uneven distribution of mobility means in particular can pose certain challenges for deprived neighbourhoods. Such areas are often characterised both by limited access to mobility means and by disadvantages caused by the road traffic of other people (see above), not rarely in proximity to the flourishing neighbourhoods of the gentrifiers. Besides the fact that the threshold for classifying neighbourhoods as 'deprived' is not easy to establish and that the related labelling of a neighbourhood can constitute a further form of stigmatisation, there seems to be an objective need for improving the living conditions, lowering environmental degradation and offering access to the means of mobility. In this case it appears to be important to identify the forms of deprivation that are typically associated with transport and traffic and to localise them in the urban space, then to develop appropriate strategies to solve these problems. Fundamentally, the adverse impacts of motorised transport in deprived urban neighbourhoods must be reduced; at the same time, actually or potentially deprived population groups must be given access to mobility and transport to enable them to overcome social and economic disadvantages of their own accord. The approach adopted in Britain for promoting social inclusion – or eliminating social exclusion – attempts to incorporate social integration more strongly in local transport development planning and in accessibility planning by ensuring mobility and participation opportunities. A general awareness of these problems and circumstances among municipal politicians, in urban and transport policy and planning will be supportive, even if deprived areas appear to be extremely difficult in terms of policy implementation, public participation, etc.
- Third, suburban areas that are approaching mature stages or phases in their lifecycle (see above) may also face serious problems in terms of social and spatial inequity, since their resilience against such problems seems to be limited, due to a lack of density, poor infrastructure and public services supply and also their often remote situation in space.

> In principle, we assume that suburban areas with a certain degree of density, infrastructure, and local quality can adapt to changing framework conditions. A number of well-known factors come to bear. First, a minimum of land-use density (population, jobs, etc.) is indispensable for acceptable public transport services, whose efficiency ensures both the urban compatibility of traffic and socially equitable access to mobility. Second, a balanced mix of uses, at least on a medium scale, is essential for (relatively) short distances and flexibility in consumer behavior with

regard to rising transport costs. Even where opportunities are provided on a small scale, a greater mix can help meet these goals; for instance, the further development of service stations into small suppliers/convenience centres. Growing transport costs may concentrate everyday activities, especially in low to medium-income households. Only if individual life worlds continue to be reflected in regional action spaces rather than in the immediate neighbourhood are narrow limits set to the acceptable spatial allocation of functions. (Hesse and Scheiner 2007, 46)

The major advantage of research approaches like the one we presented here is that based on such findings, policy and planning do not only emphasise abstract places but may develop an understanding of their complex life-worlds, including life situation, household organisation, lifestyle, etc. This would further support a shift in paradigm in mobility policy and urban planning in the sense that the specific configuration of social milieux has been added to the view of transport as a quantitative phenomenon. It means that particular emphasis is placed on accessibility for people from different social backgrounds and in different spatial situations – mostly from an overall socio-spatial perspective encompassing locational quality, social composition, accessibility and mobility options.

Action is thus needed and possible over and above merely building and operating transport facilities: aspects of mobility system design pertaining to control, organisation, fares and information have to be integrated. This is in keeping with the extension of urban renewal approaches from formal urban renewal rehabilitation to the organisational and supportive development of urban neighbourhoods as practiced in the 'Socially Integrative City' programme or participative urban redevelopment.

References

Blasius, J. and Dangschat, J.S. (1990), *Gentrification. Die Aufwertung innenstadtnaher Wohnviertel* (Frankfurt/Main: Campus).
Bonss, W. and Kesselring, S. (2001), 'Mobilität am Übergang von der Ersten zur Zweiten Moderne', in Beck, U. and Bonss, W. (eds), *Modernisierte Modernisierung* (Frankfurt/Main: Suhrkamp).
Bullard, R.D. and Johnson, G.S. (eds) (1997), *Just Transportation. Dismantling Race and Class Barriers to Mobility* (Gabriola Island: New Society Publishers).
Cresswell, T. (2006), *On the Move. Mobility in the Modern Western World* (New York: Routledge).
Department of Transport and the Regions (DETR) (2000), *Social Exclusion and the Provision and Availability of Public Transport* (London: The Stationary Office).

European Council of Ministers responsible for Urban Development (2007), *Leipzig Charter on Sustainable European Cities*, final draft (May 2).
Gray, D., Shaw, J. and Farrington, J. (2006), 'Community Transport, Social Capital and Social Exclusion in Rural Areas', *Area* 38:1, 89–98.
Handy, S., Cao, X. and Mokhtarian, P. (2005), 'Correlation or Causality Between the Built Environment and Travel Behavior? Evidence from Northern California', *Transportation Research D* 10:6, 427–44.
Heitmeyer, W., Dollase, R. and Backes, O. (1998), *Die Krise der Städte. Analysen zu den Folgen desintegrativer Stadtentwicklung für das ethnisch-kulturelle* Zusammenleben (Frankfurt/Main: Suhrkamp).
Hesse, M. and Scheiner, J. (2007), 'Suburban Areas – Problem Neighbourhoods of the Future?', *German Journal of Urban Affairs* 46:2, 35–48.
Kim, J., Pagliara, F. and Preston, J. (2005), 'The Intention to Move and Residential Location Choice Behaviour', *Urban Studies* 42:9, 1621–36.
Kitamura, R., Mokhtarian, P.L. and Laidet, L. (1997), 'A Micro-Analysis of Land Use and Travel in Five Neighborhoods in the San Francisco Bay Area', *Transportation* 24, 125–58.
Krizek, K.J. (2003), 'Residential Relocation and Changes in Urban Travel. Does Neighborhood-Scale Urban Form Matter?', *Journal of the American Planning Association* 69:3, 265–81.
Lanzendorf, M. (2002), 'Mobility Styles and Travel Behavior. An Application of a Lifestyle Approach to Leisure Travel', *Transportation Research Record* 1807, 163–73.
Long, J.S. (1997), *Regression Models for Categorical and Limited Dependent Variables* (London: Sage).
Lyons, G. (2003), 'The Introduction of Social Inclusion to the Field of Travel Behaviour', *Transport Policy* 10:4, 339–43.
Preston, J. and Rajé, F. (2007), 'Accessibility, Mobility and Transport-related Social Exclusion', *Journal of Transport Geography* 15, 151–60.
Scheiner, J. (2005a), 'Auswirkungen der Stadt- und Umlandwanderung auf Motorisierung und Verkehrsmittelnutzung: ein dynamisches Modell des Verkehrsverhaltens', *Verkehrsforschung Online* 1:1, 1–17.
Scheiner, J. (2005b), *Lebensstile, Standortbewertungen und Wohnmobilität. Analysen der Haushaltsbefragung des Projekts StadtLeben.* Raum und Mobilität – Arbeitspapiere des Fachgebiets Verkehrswesen und Verkehrsplanung 13 (Dortmund: Universität Dortmund).
Scheiner, J. (2006), 'Housing Mobility and Travel Behaviour: A Process-oriented Approach to Spatial Mobility. Evidence from a New Research Field in Germany', *Journal of Transport Geography* 14:4, 287-98.
Schweer, I.R. and Hunecke, M. (2006), 'Die Lebensstile in StadtLeben', in Beckmann, K.J., Hesse, M., Holz-Rau, C. and Hunecke, M. (eds), *StadtLeben – Wohnen, Mobilität und Lebensstil* (Wiesbaden: Verlag für Sozialwissen-schaften), 55–61.

Chapter 10
Mobility and the Promotion of Public Transport in Johannesburg

Ursula Scheidegger

Introduction

Transport is inherently linked to the city; it is not only a critical aspect of sustainable cities but also a defining aspect of inequality (Blowers and Pain 1999, 279). Social and economic conditions of people determine transport demands and social class influences mobility, accessibility and choices. Policies and infrastructural development are designed to accommodate the requirements of dominant social groups offering or limiting opportunities of individual commuters and often travel patterns do not reflect actual needs of travellers. Means of transport are not as important as the purpose of a journey, transport is a derived demand; hence facilities are often rudimentary and ignore the experiences of the travellers (Hamilton and Hoyle 1999, 58). People use inadequate travel provisions because they have no alternative (Vasconcellos 2001, 207). In addition, pedestrians and non-motorised transport are almost entirely neglected despite the fact that walking is an essential if not the predominant means of transport in cities, does not contribute to environmental pollution and allows for human interactions in the public realm (Vasconcellos 2001, 111). Transport networks bring people together but also divide communities; usually pedestrians have to make detours to a pedestrian bridge or underpass in order to get safely across the freeway or railway tracks and people living closest to the freeway often belong to the social class that can the least afford a car (Hamilton and Hoyle 1999, 52).

Transport in Johannesburg – one of the host cities of the soccer World Cup in 2010 – is no exception and traffic is a factor determining the quality of life. Transport renders Johannesburg dangerous, reinforces social inequalities, contributes to high levels of stress for travellers, is a critical producer of pollution, traffic congestion costs millions in terms of unproductive time and waste of fuel and accidents are a threat to human life. Johannesburg's current transport system is a legacy of apartheid adapted to the needs of a privileged minority and offering inadequate transport arrangements for the majority of commuters.

After the first democratic elections in 1994, the newly elected government had to address the inequalities inherited from the past in order to provide for satisfactory living conditions of the entire population, which included transport provisions and related aspects of mobility, accessibility and choices. This chapter discusses changes in the transport sector; it starts with a historical overview followed by governmental strategies after 1994. Beside the aspect of more equity in public spending, intolerable levels of road congestions and the prospect of hosting the Soccer World Cup in 2010 are further factors contributing to the urgency to address the public and private transport situation and promote public transport. In addition, the gains of investment in the transport infrastructure in the wake of the Soccer World Cup last beyond 2010. The recapitalisation of the taxi industry – a semi-formal mini-bus service – is used as an example of governmental programmes addressing urban transport since the majority of commuters rely on taxis. Furthermore, taxis would be one transport option to the World Cup matches. This chapter argues that it is extremely difficult to change the attitudes of people, in particular of the middle class, towards public transport for several reasons: perceptions of public transport are influenced by the low quality of the current system, the private car is considered the solution of all transport problems despite the fact that the majority of the population depends on public transport, the prevailing perceptions of public spaces as locations of disorder, danger and vice, which is daily reinforced by intolerable levels of crime and finally because of the ways the different population groups relate to each other. References to the different population groups are done in the terminology commonly used in South Africa.

Historical Context

The discovery of gold in the Witwatersrand in 1886 marks the beginning of Johannesburg and already in 1891, a system of public transportation was introduced, the horse-drawn tram (Beavon 2004, 57). Electrical trams started to operate in 1906, expanding their services to the suburbs of the fast-growing city (Beavon 2004, 90). However, the ever-expanding low-density residential areas of the white population worked against a viable public transport system, in addition, car ownership rapidly increased. In 1948, the trams were eliminated and were only partly replaced by buses, which together with the expansion of the road network further encouraged the use of private vehicles (Beavon 2004, 161).

From the early beginnings of Johannesburg, native locations were established for the non-white population; the Land Act of 1913 limited property ownership of Africans to three freehold areas: Sophiatown, Kliptown and Alexandra. The introduction of the infamous Group Areas Act in the 1950s not only forced the non-white population to the periphery of the city far away from work opportunities, but also exacerbated the already inadequate transport situation of the respective population. Besides the long travelling distances between home and work, the provision of trains and buses was not sufficient; hence people had to get up as early as four in the morning to catch a train or bus in order to arrive at work in time. In addition, workers did not earn enough to pay for their transportation at market rates and the state had to help out with subsidies amounting to 1.3 billion Rand in 1988/89, an amount that does not include subsidised school transport. From the late 1970s onwards, mini-bus kombis – the so-called taxis – started operating; however, it was difficult to obtain permits and many taxis operated illegally. Over the following ten years 30 per cent of African commuters started to use taxis and between 1984 and 1989, thanks to the taxi industry, the average daily travelling time of Africans was reduced from three to two hours (McCaul 1991, 218–29). The taxi industry was growing fast and provided business opportunities for the black population; usually the owner of a taxi company employs drivers, there is little control of driving skills and the unregulated industry lacks any stipulations regarding salaries and social benefits of drivers. The booming industry soon resulted in an oversupply and competition for routes and violent confrontations at the expense of commuters regulated shares in the market of individual taxi companies. Furthermore, the industry is not profitable; as a result, the maintenance of taxis is minimal, overcrowding increases profits and despite legislation limiting the maximum number of passengers, transgressions are often a possibility for taxi drivers to prop up their meagre salaries and the faster one drives the more people one can transport. All these factors contribute to the high number of taxis involved in accidents. The current vehicles most often used by the taxi industry are Toyota buses, imported as second hand cars from Japan, where the vehicles were used on industrial sites – for short routes and a small number of passengers. Hence the vehicles are unsuitable for the way they are used in Johannesburg – long routes and as many passengers as possible (information provided by Ncanana). Since there is no coordination of the taxi industry, routes go from the townships to the city centre and from there to the different locations of employment in business districts or residential areas, often extending travel time and cost, because direct routes at the periphery or diagonally across the city would reduce the travel distance considerably. Along these taxi routes a vibrant informal market for goods such as drinks, sweets and food developed, which together with taxi ranks contributes to disorder and competition for space in business districts and residential areas.

Political Transition and Democratisation

The political unrest at the end of the 1980s and the uncertainty during the transition contributed to low levels of law enforcement, which extended to South Africa's roads. Together with the perceived right to unlimited mobility of car drivers, public planning in favour of motorists at the expense of pedestrians contributed to a situation where the weakest segments of society experience the highest risks while using public space and transport. Hence, the priority of motorised transport resulted in road development inherently dangerous for the majority of the population (Vasconcellos 2001, 205). Even worse, the power of motorised transport and poor law enforcement contributed to a feeling of impunity among motorists which reinforced bad traffic behaviour (Vasconcellos 2001, 192–5). In 1994, the new government inherited a transport situation completely biased in favour of motorised transport and an immense degree of lawlessness that resulted in one of the highest road fatality rates in the world, exacerbated by the rapidly increasing number of cars on South Africa's roads. In addition, inherited public transport arrangements from the apartheid era were poorly developed, inefficient, unreliable, badly serviced and dangerous. Routes often bypassed central areas bringing workers from the townships directly to industrial sites (Mabin 1991, 34) making it unfeasible to upgrade and improve these services.

In highly unequal societies such as South Africa, dominant social groups not only have choices but also control change and with regard to transportation, tend to neglect inadequate travel conditions of the urban underclass (Hamilton and Hoyle 1999, 65); a poor transport infrastructure is a characteristic of marginalised areas, where structurally irrelevant people live (Mooney 1999, 81). Furthermore, poor people have limited perceptions of citizenship, rights and duties and often lack the means to influence state policies (Vasconcellos 2001, 213). Hence, the state has an important role in managing traffic and negotiating between the different population groups and stakeholders, a conflictual process because the various population groups are affected by public policies in very different ways. Beside accessibility and safety, fluidity, cost and environmental considerations are important aspects of transport management in a context where spatial, topographic and built structures set limitations on development options. It is critical to be aware of the different political forces that command public policy of urban traffic and integrated mobility management in order to do justice to all segments of society (Vasconcellos 2001, 63–74).

The *White Paper on Transport* in 1996 was a first attempt by the democratically elected government to develop strategies and intervene in the traffic situation. The document stipulates the importance of increasing road safety, enforcing discipline and the law and promoting the use of public transport over private car travel with the goal to achieve a ratio of 80 to 20 per cent between public and private transport (White Paper 1996). Within the limited context of this chapter it is impossible to discuss all aspects of development in the urban public transport sector, which includes trains, fast-track and other buses and the taxi industry. The chapter focuses on the taxi industry and its recapitalisation programme, because taxis are the largest sector of public transport and secondly, as a semi-formal structure, the taxi industry does not enjoy any governmental subsidies in contrast to other means of public transport and road development, which conflicts with conceptions of equality in public spending of the democratic state. In addition, more justice and equality in the allocation of public resources and provisions of services also comprises the principle of equity. With regards to traffic, equality is reflected in provisions of services to all neighbourhoods whereas equity considers the specific characteristics of individual commuters, because income, gender, age or physical challenges might interfere with the ability to enjoy provided services (Vasconcellos 2001, 245).

The current development of the public transport sector is an integrated part of governmental strategies to address transport needs during the Soccer World Cup in 2010. Beside the improvement of the existing modes of public transport and the construction of a fast train – the 'Gautrain' from the Oliver Tambo International Airport in Johannesburg to Tshwane (previously Pretoria), the business district in Sandton and the city centre of Johannesburg – the different municipalities have provided plans and suggestions of how to organise transportation within their jurisdiction, which are currently under evaluation. A special task team coordinates and monitors this process and the ensuing development of transportation strategies, plans and their implementation (information provided by Ncanana and Seedat).

The Taxi Recapitalisation Programme

Approximately 120,000 taxis (according to government estimate) to 200,000 taxis (according to operator estimate) operate in South Africa. They are on average more than ten years old and hence not suitable for public transport operations; nevertheless, they carry 14 million people each day, this is over 60 per cent of public transport users and about 30 per cent of all workers. 18 per cent of all households spend more than 20 per cent of their income on public transport, 31 per cent of all households more than 10 per cent. The estimated value of the industry is 12 billion Rand a year (Department of Transport 2005, 23; Statistics South Africa 2005, 70–77; *Saturday Star* 2007, 2). Taxi drivers are hated by commuters as well as by motorists for their disrespect of traffic regulations and the resulting threats to passengers and other road users, which is exacerbated by occasional resort to violence during confrontations and disputes. Law enforcement is weak and was for a long time non-existent; taxi drivers just ignored traffic fines. Today, the impounding of taxis for a day or two is a more effective strategy to bring perpetrators to book because a one-day's loss of income hurts and directly affects the individual driver.

The main objectives of the recapitalisation programme are the improvement of the quality and the better monitoring of services, stricter law enforcement, affordable operations and the priority of safety and convenience of travellers. One aspect of the taxi recapitalisation programme is the technicalities and the legal frame, the other is the reactions and acceptance of the various stakeholders in the industry, which are affected by the changes in very different ways. The programme has been developed in a transparent way, allowing for participation by the different stakeholders; however, the qualification and skills of the taxi drivers are not negotiable and better control will minimise the options of illegal drivers. In addition, the opening of the industry to new businesses reduces the quasi monopoly of individual taxi syndicates (operators). Toyota, currently the main provider of vehicles to the taxi industry is affected, because new companies will enter the market. Rumour has it that Toyota provided taxi drivers with food while they were blocking part of the main transit roads in protest to the recapitalisation. Commuters are the main beneficiaries, since the programme improves their safety and reduces cost of transportation. Finally, new stakeholders will enter the industry, namely financial institutions and insurance companies.

The New Taxi Vehicle introduced in the recapitalisation programmes has to comply with compulsory safety that includes technical requirements such as tyres or brake systems but also regulations such as maximum speed and number of passengers. An Electronic Monitoring System (EMS) has to be installed controlling among others mileage, fuel, number of stops for passengers, excessive braking and speeding. In addition the EMS indicates when regular inspections and maintenance are due and prevents unauthorised use, the driver has to log in with a valid chip on a valid driver's licence and access the log file, otherwise, the vehicle will not start. In this way, there is also control of the compliance with legal stipulation regarding working hours of drivers. There are three different categories of new taxis catering for a maximum of 16, 28 or 35 passengers and all vehicles with an operating license are branded in order to be identified by commuters, all are white in colour with the national flag on the side and an aluminium plate with the provincial coat of arms and the route number. Current legal operators get an allowance of R50000 for the scrapping of one old vehicle towards the purchase of a new. In addition, the municipality determines public transport routes and stops – not allowing for stops in-between – current taxi routes will be supplemented by diagonal and peripheral routes and dedicated road space will be reserved for the exclusive use by public transport. An integrated ticketing system, also operated by the EMS, allows for cashless travel and links the different modes of public transport. Finally, in order to ensure quality services the municipality collects the revenue and pays operators per vehicle kilometre. Semi-formal operators are drawn into the system; legalised services have to operate within the municipality network under the terms of the integrated plan and the control by the EMS (information provided by Ncanana and Seedat; Department of Trade and Industry 2000, Department of Transport 2007).

The taxi industry was not only included in the development of the new programme, taxi drivers and operators will get training for the operation of the Electronic Monitoring System in addition to training for drivers to improve driving skills and the provision of services to commuters. However, taxi drivers fear for their jobs; not only do many taxi drivers lack valid licenses, there are concerns that a new fleet of bigger vehicles will require fewer drivers. Taxi operators will have to comply with legal requirements, which impacts on revenue and in a more competitive market, control and hence power are reduced. A first step was to compel the industry into taxation; a tax amnesty expiring end of April 2007 was the last chance to legalise the business or face legal consequences. Despite the exploitative employment structure in the taxi industry, taxi operators and drivers team up in order to contest the recapitalisation programme. They lack popular support because of the industry's reputation; nevertheless, their actions are disruptive and often turn violent and affect the wider public. Since law enforcement is weak, daily demonstrated by the violation of traffic regulations by the taxi industry, there are doubts whether the government will be able to fully implement the recapitalisation programme, despite the assurance of officials not only to go ahead but also to rigorously prosecute non-compliance. This is a difficult environment for winning support in favour of public transport (information provided by Ncanana and Seedat).

Popular Attitudes towards Public Transport and Public Space

The car is important in South Africa, not only as a means of transport but also as a status symbol. The car increases mobility and thus access and opportunities and any changes regarding the use of cars are also a profound challenge to the lifestyle of the urban middle class that has evolved around the car (Blowers and Pain 1999, 277). Nevertheless, public transport will improve as soon as the middle class starts using it. Speed and freedom are other powerful arguments in favour of the car, in contrast to public transport it allows for direct travel from door to door at any time, it provides privacy, a place to be alone and one can choose fellow travellers; hence the access to a car is equated with personal liberation. However, car users navigate in an impersonal space, human interactions are reduced to a minimum and anti-collective attitudes develop. In addition, the car represents power; this is not only reflected in the uneven allocation of public resources in favour of powerful social groups – only 26 per cent of households in South Africa have access to private cars and 79 per cent of South Africans do not have driving licences (Department of Transport 2005) – as a status symbol the car creates and nurtures inequality and thus a fancy, elegant, expensive, huge or fast car is for some people a vital expression of their personality. Car ownership also reflects gender disparities: the majority of car owners are men, they monopolise the use of the car in one-car households, men often drive fancier cars than their female companions and there is an obvious relationship between power and speed (Hamilton and Hoyle 1999, 60–64). Irrespective of the cost involved regarding car ownership and increasing road congestion, South Africans are socialised into the perception that the car is desirable and the solution of all transport problems. The obsession with cars is a difficult environment for the change of perceptions and attitudes towards public transport in particular because of the quality of current public transport services.

Public spaces are important institutions of cities and public transport is a way of negotiating and using public space. In contrast to the limitations of the private space where one is surrounded by family and friends and relations are reliable, public spaces offer opportunities of new experiences. However, the growth of cities and the anonymity of urban space also allowed for the accommodation of different lifestyles and deviant individuals and groups and increasingly the private space became associated with virtue and security and the public space with disorder, vice and danger (Sennett 1976, 16–24). In South Africa, apartheid regulated the use of public spaces by restricting and controlling access not only profoundly influencing ways of interactions but also perceptions of other population groups. In post-apartheid South Africa, access and utilisation of public space continues to be monitored by various social forces and current levels of crime prevent social integration. Despite constitutional rights and the abolishment of the Group Areas Act, de facto not everyone enjoys equal rights and access to public space is often determined by assumptions based on decency and moral worth, hence public space is open for those who deserve it and are considered to be capable of acting in a responsible and predictable way (McDowell 1999, 111–13).

Insecurity perpetuates racial stereotypes and fosters the proliferation of gated communities and 'fortified' shopping malls (McLaughlin and Muncie 1999, 122). The privatisation of public space can provide a sense of stability and confidence in an environment of conflicting rights related to incompatible social and economic differences, but it also produces a form of exclusive citizenship. In these security clusters, fear is a function of security mobilisation and creates its own demand that is less about safety and more about insulation, because most of the crime and violence happens in the townships and not the suburbs (Davis 2003, 202). Nevertheless, in gated communities people constrain themselves and their social interactions, walls produce extreme forms of insular subjectivity, normalise paranoid attitudes, demonise the other and generate chronic anxiety. People choose to remain strangers to each other (McLaughlin and Muncie 1999:120–25). These 'urban fortresses' are anti-pedestrian and restrict opportunities of interactions reaching across social divisions in the public realm; they are repressive and their motive is control. Furthermore, the use of the car and the design of freeways and urban transit roads not only inhibit personal interactions but also allow the middle class to navigate the city without encountering the urban underclass or being confronted with township realities. Finally, the privatisation of the public realm and increased surveillance are not only responses to insecurity and crime but also manifestations of the economic deregulation, the commoditisation of the commons and the devaluation of non-market entitlements (Davis 2003, 203).

South Africa not only has one of the highest road accidents and crime rates; it also is one of the countries with the highest alcohol consumption and has the highest Gini coefficient, statistics that reflect social pathologies. The revival of public space is only possible without social exclusion and the denial of basic rights. Hence confidence in public space and the willingness to use public transport is also inherently linked to the state's capacity to address social ills; disorder and insecurity cannot be understood in isolation from inequality (Mooney 1999, 88). The continuous presence of violence and death foster indifference and insensitivity towards the value of human life and the spatial and social distance of the middle class from disadvantaged population groups prevents the development of a sense of social responsibility and an acknowledgement of the interdependence between the different social groups (Kalati and Manor 1999, 122). There are people who have nothing to lose, a factor that encourages high risk behaviour. In addition, alcohol consumption is part of sociability and one drink too many is socially accepted; violence is often alcohol related and speeding and drunk driving are the main contributors to the carnage on South Africa's roads.

What is the way forward? Experiences of other cities tell us that more and wider roads do not increase the flow of traffic but attract more cars and motorised households tend to make three times more trips than public transport users (Carley 2001, 5). Hence more space for private transport is not an option to address the transportation problem; road congestion is the worst possible outcome for everyone (Vasconcellos 2001, 192). Strategies need to promote a change in consciousness and behaviour, in particular more respect for pedestrians and a more positive attitude towards public transport. The promotion of safety is imperative as a right of everyone and the preferential treatment of pedestrians shifts the focus from the importance of the car to the superiority of humans in any traffic situation (Vasconcellos 2001, 281), well reflected in the governments terminology change from transport-flow to people-flow. However, it also needs incentives that encourage individuals and households to take a more realistic account of the collective cost of their behavioural choices (Downs 2003, 264).

Governmental Strategies Promoting Public Transport

The Strategic Public Transport Network is an integrated programme addressing the current short-comings of public transport and creating incentives to increase the number of public transport users. It starts with the upgrading and development of existing services, the New Taxi Vehicles or more modern trains and buses and the introduction of express trains and express buses are supplemented with more reliable and efficient services. Preferential treatment of public transport by marking road space for exclusive use ultimately aims reducing travel time in comparison to private transport and the priority of humans in any traffic situation limits the power of motorists. However, while technical improvement and better management are feasible, the respect of regulations, whether the superiority of humans or preferential treatment of public transport depend critically on the capacity of law enforcement (information provided by Ncanana).

Further strategies promoting attitudinal changes are the different programmes raising awareness for the advantages of public transport, starting with the Gauteng Department of Public Transport's annual promotion of Public Transport Month in October and its related activities – a highlight is the celebration of Car Free Day. This will be complemented by the planned demarcation of car free zones and financial incentives increasing cost of private transportation for example the introduction of more toll-roads and a congestion fee. Transport challenges should be turned into a Gauteng citizenry issue, not only by raising awareness of public transport as an alternative but also by pointing at the environmental benefits, informing the public about strategies to reduce congestion, encouraging cooperation towards sustainable mobility and promoting a healthy lifestyle and the positive consequences of walking and the use of non-motorised transport (Gauteng Department of Public Transport 2006; Department of Transport 2007). The campaigns during Public Transport Month and Car Free Day are accompanied by temporary measures for example the reservation of one lane on the Joburg–Tshwane Freeway for high occupancy vehicles (three or more passengers), the closure of roads for private transport, the launch of commuter organisations and public meeting points for car pooling. A concerted media campaign informs about the programme and related activities; on Car Free Day, the use of public transport by provincial government officials is well documented in the evening news and should serve as an example. Generally, public transport enjoys a fairly high media presence (Gauteng Department of Public Transport 2006; Department of Transport 2007).

Unfortunately, the built environment limits these efforts not only because of the low population density in the suburbs but also because of the spatial distance between homes, work, shopping and leisure. In areas with a diversity of functional uses – entertainment, living, shopping and offices within walking distance the number of people in the street provides a degree of security and social control (Jacobs 1961, 245). Since land use changes along public transport routes, mixed functional areas are planned for example the hubs at the different stations of the Gautrain. Yet, the question remains whether alienated populations are willing to share travel space considering the impossibility of choosing fellow travellers. Nevertheless, public spaces provide opportunities of meaningful relations and in the context of Johannesburg also of social interactions. Finally, safety is critical for the promotion of public space and public transport, on the other hand, public space gets safer the more people used it.

Conclusion

Johannesburg's current transport situation reflects the legacies of the past in terms of spatial urban development, biased approaches to mobility, the perception of public spaces, safety and the quality of social interactions between the different population groups. Governmental strategies aim at improving transportation and are driven by more equity in public spending, increasing road congestion and transport needs during the soccer World Cup in 2010. Integrated programmes not only include the upgrading, expansion and improvement of the public transportation network but also promote attitudinal changes of private transport users. However, the confidence in public spaces and the willingness to use public transport is critically linked to the state's capacity to address incompatible social and economic differences and the resulting instability and insecurity, to implement and enforce transport policies and effectively prosecute transgressions. Safety is also the biggest challenge for the upcoming World Cup. In a safe environment, where the different population groups are also willing to engage with each other, public transport could serve as a space of integration not only by offering mobility but also by connecting people.

References

Ajam, K. (2007), 'The Taxi Industry in Numbers', *Saturday Star*, 2 June.
Allen, J., Massey, D. and Pryke, M. (eds) (1999), *Unsettling Cities: Movement/Settlement* (London: Routledge).
Beavon, K. (2004), *Johannesburg The Making and the Shaping of the City* (Pretoria: University of South Africa Press).
Blowers, A. and Pain, K. (1999), 'The Unsustainable City?', in Pile, S., Brook, C. and Mooney, G. (eds), *Unruly Cities? Order/Disorder* (London: Routledge).
Carley, M., Jenkins, P. and Smith, H. (eds) (2001), *Urban Development and Civil Society* (London: Earthscan).
Carley, M. (2001), 'Top-Down and Bottom-Up: The Challenge of Cities in the New Century', in Carley, M., Jenkins, P. and Smith, H. (eds), *Urban Development and Civil Society* (London: Earthscan).
Davis, M. (1990), "Fortress L.A'. from City of Quartz: Excavating the Future in Los Angeles (1990)', in LeGates, R.T. and Stout, F. (eds), *The City Reader* (London: Routledge).
Downs, A. (2003), 'The Need for a New Vision for the Development of Large US Metropolitan Areas', in LeGates, R.T. and Stout, F. (eds), *The City Reader*, 3rd edn (London: Routledge).
Hamilton, K. and Hoyle, S. (1999), 'Moving Cities: Transport and Connections', in Allen, J. Massey, D. and Pryke, M. (eds), *Unsettling Cities: Movement/Settlement* (London: Routledge).
Jacobs, J. (1961), *The Death and Life of Great American Cities* (London: Jonathan Cape).
Kalati, N. and Manor, J. (1999), 'Elite Perceptions of Poverty: South Africa', *Institute of Development Studies Bulletin* 30:2, 117–26.

LeGates, R.T. and Stout, F. (eds) (2003), *The City Reader*, 3rd edn (London: Routledge).
Mabin, A. (1991), 'The Dynamics of Urbanisation since 1960', in Swilling, M., Humphries, R. and Shubane, K. (eds), *Apartheid City in Transition* (Cape Town: Oxford University Press).
McCaul, C. (1991), 'The Commuting Conundrum', in Swilling, M., Humphries, R. and Shubane, K. (eds), *Apartheid City in Transition* (Cape Town: Oxford University Press).
McDowell, L. (1999), 'City Life and Differences: Negotiating Diversity', in Allen, J. Massey, D. and Pryke, M. (eds), *Unsettling Cities: Movement/Settlement* (London: Routledge).
McLaughlin, E. and Muncie, J. (1999), 'Walled Cities: Surveillance, Regulation and Segregation', in Pile, S., Brook, C. and Mooney, G. (eds) (1999), *Unruly Cities?* (London: Routledge in association with The Open University).
Mooney, G. (1999), 'Urban "Disorders"', in Pile, S., Brook, C. and Mooney, G. (eds), *Unruly Cities?* (London: Routledge in association with The Open University).
Pile, S., Brook, C. and Mooney, G. (eds) (1999), *Unruly Cities?* (London: Routledge in association with The Open University).
Swilling, M., Humphries, R. and Shubane, K. (eds) (1991), *Apartheid City in Transition* (Cape Town: Oxford University Press).
Vasconcellos, E.A. (2001), *Urban Transport, Environment and Equity* (London: Earthscan).

Government Sources

Department of Trade and Industry (2000), *Taxi Re-capitalisation Project.*
Department of Transport (2005), *Key Results of the National Household and Travel Survey.*
Department of Transport (2006), Pamphlet October Transport Month.
Department of Transport (2007), *Strategic Overview and Key Policy Developments: 2007/8-2009-10.*
Gauteng Department of Public Transport, Roads and Works (2006), *Slideshow Public Transport Month.*
Government of the Republic of South Africa (1996), *White Paper on Transport* (Pretoria: Department of Transport).
Statistics South Africa (2005), *General Household Survey, July 2005.*

Additional Information

M. Ncanana, Gauteng Provincial Department of Transport, Johannesburg.
I. Seedat, Director Public Transport Strategy, Department of Transport, Tshwane.

Index

academia/academic 1, 2, 28, 39–40, 44, 77, 92, 145, 155, 168, 198
access 2, 8–9, 11–15, 18, 20–21, 31, 33–7, 60, 75, 77, 80, 81, 86–7, 90, 103–4, 121, 123, 146, 155, 167, 179, 187, 189, 194, 198, 200–201, 203–4, 212–14
accessibility 12, 17–20, 27, 99, 140, 189, 194, 198, 204–5, 207–8, 210
  social 16, 20, 102
actors 9, 11, 14, 52, 121, 123, 156, 165, 166, 167, 168
Africa 1, 148, 172, 210–11, 213
age 9, 13, 18, 32, 38, 65–6, 91, 107, 110, 112, 114, 125, 137, 139, 171, 190, 192, 197, 211
Albrow, M. 14–7, 165
Allan, G. 169, 177
apartheid 207, 210, 214
appropriation 12, 15, 27, 34–5, 41, 87, 154
Asia 14, 52, 172
asylum seekers 10
attractiveness 16, 64, 65, 196
Augé, M. 84, 147
Austria 149, 154, 156
Axhausen, K.W. 99, 101, 102, 105, 106, 115, 135, 147, 169

Bangladesh 1
Bauman, Z. 28–9, 36, 37, 75–6, 78–85, 88–92, 165, 167
  'escape' 76, 81–4, 88
  'exit' 76, 81–4, 88
  'mismeetings' 76, 83–4, 88, 91
Beck, U. 10, 28, 54, 59, 67, 93, 100, 104, 116, 147, 156
Bergman, M.M. 1, 7–9, 14, 30, 32–3, 99
Bernardi, F. 54, 56, 57
Blossfeld, H.-P. 51, 53, 55–60, 62–5
Bourdieu, P. 15, 28, 32, 34–5, 43, 45, 85, 123
Breen, R. 54, 59, 67

Buchholz, S. 51, 63–5

Canada 55
capital 9, 11, 13–15, 20, 29, 30, 34–8, 41, 45, 51, 56, 60–61, 65–6, 76, 85–90, 92–3, 101, 122–4, 136–8, 140, 166, 170, 175
  cultural 14–15, 20, 34–6
  economic 9, 35, 87
  human 56, 60, 61
  mobility 13, 138
  network 29, 35–6, 41, 45, 76, 85–90, 92–3
  social 14, 38, 88, 90, 92, 101, 122–4, 136–8, 140, 170
  bridging 123–4, 136–7, 140
Cass, N. 18, 20, 33, 35–6, 40, 101, 121, 139, 169
Castells, M. 28, 30–31, 52, 85–7, 89
class 2, 8, 10, 13, 18, 35, 38, 54, 60, 67, 81, 146, 149, 154, 166–7, 169–71, 174, 180–81, 187, 207–8, 213–15
communication 1, 12, 14, 20, 27, 29–30, 35, 40–41, 44–5, 51–2, 66, 75, 79, 86–7, 89, 99, 101, 104–5, 121, 138, 165–71, 179
  email 11, 35, 87–8, 105, 108
  face-to-face 86–7, 89–90, 103, 105–6, 108, 110, 114, 168–9, 193
  fax 11
  letters 11, 88
  mobile phone 11, 33, 35, 41, 108, 154
  telephone 11, 41, 87–8, 105, 108, 123, 153–4, 174
  text messages (SMS) 88, 105, 108
communities 19, 88, 169, 196, 207, 214
  gated 82–3, 92, 214
commuting 31, 121–32, 134–41, 153, 157–8
  emotional support for 126, 128, 130–2, 134, 136

competence 12, 15, 34, 87, 155, 168, 180
connectedness 16, 28, 29
Cook, Thomas 41, 42
Cresswell, T. 27, 37–9, 187–8
Czech Republic 55, 61

Denmark 55, 63, 66
deregulation 52, 214
differentiation 2, 7, 37, 85, 102, 176, 202
   social 10, 15, 20, 32, 116, 157
displacement 10, 11, 27, 37, 39
dynamics 1, 2, 7–12, 21, 51, 66, 104–5, 122, 136, 140, 152, 203
   identity 10
   mobility 11
   social 21
   societal 7
   spatial 51, 66
   urban development 203

Eastern Europe 52, 57, 58, 60
East Germany 177, 178, 179
education 10, 17–18, 31–2, 35, 54, 56–61, 63, 65, 101–2, 107–8, 110, 112, 114–15, 126, 128–9, 131–2, 134, 137–9, 166, 171–4, 176, 179, 189, 192–3, 196, 198
   educational level 138, 171–5, 177, 179–80
   educational attainment 2, 9, 61, 171, 179
employment 14, 18–19, 54–67, 105, 174, 179, 192, 209, 213
Estonia 55, 61
ethnicity 13, 18, 38, 188, 196
Europe 38, 54, 56–8, 60, 146, 189
Europe 64, 65
   European Commission 168
   European Council of Ministers 203
   European Union 168

family 11, 14, 18, 20, 43, 55, 57–8, 60–63, 66, 86–7, 89, 92, 103–4, 108, 112, 125, 128–9, 131, 133–4, 137–40, 145–6, 149, 153, 155–7, 169, 175, 177, 193, 196, 197, 214
   extended 103
   nuclear 103
flexibilisation 58–9, 64–7, 149, 189

flexibility 54, 57–8, 61–6, 103, 116, 204
flows 1, 28, 30–31, 38, 86, 201
France 55, 150
Frei, A. 99, 105, 106, 121, 135, 169
Frello, B. 27, 36, 38–9
friendship 20, 86, 89, 92, 169, 197

gender 9, 10, 13, 17–20, 32, 38–9, 51, 54–7, 59, 60–63, 66–7, 80, 103, 107, 110, 127, 130–31, 138, 171, 177, 178–9, 192, 196–9, 211, 213
gentrification 188, 203–4
Germany 17, 18, 51, 55–6, 103, 152, 156, 158–9, 166, 170–71, 174–80, 188
   Cologne 188, 190–92, 198, 201–2
Giddens, A. 54, 165
GLOBALIFE 51–2, 54–6, 58–60, 62–7
globalisation 1, 15, 28–9, 41, 45, 51–67, 79, 91, 99, 101, 165–7, 177, 180, 203
Goffman, E. 78, 89
Goldthorpe, J.H. 8–10
Graham, S. 28, 30, 31, 41
Grieco, M. 2, 16–17, 20, 102, 104, 177
Grounded Theory 148, 150
Gusfield, J.A. 76–8

Hall, M.C. 14, 147, 152
Hamilton, K. 40, 207, 210, 213
Handy, S.L. 16, 102, 194, 202
Hannerz, U. 147, 165, 169
Harvey, D. 28, 30, 89
health 9, 17, 19, 31
Hesse, M. 187, 190, 205
Hine, J. 2, 30–31, 102, 104
Hofäcker, D. 51, 64, 66
Hofmeister, H. 55, 60, 62–3, 66
Holden, E. 32, 33, 37
homelessness 92, 151
housing 81, 145, 155, 187–8, 190–92, 201–2
Hoyle, S. 207, 210, 213
Hungary 55, 61

immigrants 38–9, 196
immobility 34, 37–9, 81–2, 90–91, 152, 165
income 2, 9–10, 14, 17–19, 27, 56–7, 61, 63, 104–7, 110, 114, 189, 191–2, 199, 202–3, 205, 211

high 18–19, 191
household 106, 192, 199
low 17–19, 63, 104, 191
middle 114
India 1
individualisation 1, 67, 149, 156
information and communication technologies (ICTs) 20, 51, 52, 66, 86, 88, 104
infrastructure 12, 14, 19, 30–31, 40, 145, 154, 188–91, 201, 203, 204, 208, 210
interconnectedness 7, 28, 35, 38, 52, 77, 165
interfaces 86, 146, 155–6
internet 1, 10–11, 75, 87, 150, 153–4
interstices 146, 150, 152–4
Ireland 55, 58
Iron Curtain 52, 61
Italy 55, 159

Jacobsen, M.H. 75, 79, 82, 85
Japan 172, 209

Kaufmann, V. 2, 3, 7–8, 10–12, 14–15, 27, 34–5, 87, 100, 121, 138–9, 145–7, 153, 157, 169
Kenyon, S. 16, 20, 139
Kesselring, S. 34, 116, 121, 188
knowledge 12, 14, 63–4, 77, 79, 168–9, 202

Larsen, J. 11, 41, 85–6, 88–9, 92, 99, 101–5, 169–70
Le Breton, E. 2, 121, 139
leisure 11, 18, 20, 31, 102, 108, 116, 123, 145, 147–8, 153, 156–7, 192–4, 196–7, 199, 201, 203, 216
Löfgren, O. 147, 150, 151, 152
Löw, M. 29, 32, 101

Manderscheid, K. 14, 27, 30, 32–3, 35, 43
marginalisation 60–62
market 53–67, 156, 189, 191–2, 202
    housing 191–2, 202
    labour 53–67, 156, 189
Marvin, S. 28, 30–31, 41
Marx, K. 8, 37, 87
Massey, D. 28–30, 32, 75

Mau, S. 165–6, 169, 176, 180
Merkel, I. 147, 151
metaphor 37, 39, 75–93
Mexico 55
migration 1, 2, 11, 36, 88, 104, 137, 147, 167, 201
Mills, M. 53, 54, 57, 59, 60
Mitchell, F. 2, 30, 31
modernisation 1, 104
modernity 1, 37–8, 45, 75, 79–80, 83–4, 88, 91–2, 123, 137, 140, 145, 149
motility 3, 7, 11, 12–16, 20, 29, 34–6, 41, 45, 87, 122, 138–9, 146, 169–70
multilocality 145–59

narrative 11, 38, 42, 150
networked space 28, 30
networks 8, 12–14, 16, 20, 30, 32, 35–6, 55–6, 78, 85–8, 92, 99–105, 108, 110, 114, 116–17, 123–5, 131, 135–8, 140–41, 150, 153, 155, 167–72, 175–6, 180, 203, 207
    closed 123, 137
    egocentric 103, 108, 110, 116, 117
    family 103
    metaphorical 78
    social 8, 13–14, 16, 20, 35–6, 86–8, 92, 99–116, 125, 135–8, 141, 155, 165, 167–70, 176, 180, 203
    transnational 168–9
    transport 207
Norway 55, 66
Nowicka, M. 28, 165, 167, 169

OECD 37, 51, 55, 64, 166, 172
Ohnmacht, T. 1, 7, 14, 99, 101–2, 105, 116, 121, 169

Peters, P. 27, 29, 40, 41, 42
Poland 55
postmodernity 75, 79, 103
poverty 2, 9, 17, 81–2
power 3, 8–9, 13–14, 19, 20, 27–8, 30, 34, 36–42, 44–5, 53, 56, 75, 78, 82–3, 85, 88, 93, 151, 155, 210, 213, 215
    social 28, 38, 41, 45
power relations 28, 30, 36–8, 40–41, 45, 53

prestige 8–10, 13
Preston, J. 17, 20, 189
privatisation 52, 154, 214
Putnam, R. 38, 88, 124

Rajé, F. 2, 16–17, 19–20, 189
relationships 40–41, 52, 54, 59, 67, 84, 102–3, 108–116, 121–4, 126, 136–8, 140, 166, 168–71, 174, 176
  long-distance 103, 168
  retirement 55, 64–7
  early 64–6
Richardson, A.J. 21, 27, 30, 42
Rigney, D. 77, 79, 85, 91
Rolshoven, J. 147–8, 151–4

Sassen, S. 30–31, 177
Scheiner, J. 32, 101, 187, 190, 192–4, 198, 202, 205
sedentarism 38–9, 80, 82
segregation 10, 140, 188, 190, 194, 201, 203
Sennett, R. 38, 147, 155, 214
settledness 145, 147, 150–53
Sheller, M. 10, 27, 29, 100
Shove, E. 27, 35, 41, 43
Simmel, G. 89, 100, 137
Simó Noguera, C. 56–7, 63
skills 12, 14, 34–5, 41, 45, 86–7, 90, 168–70, 209, 212
Sklair, L. 165, 167–8
social change 1, 9, 14, 81, 148, 158–9
social contacts 35, 99, 101, 103–5, 108, 111–13
social exclusion 8, 10, 15–20, 37, 87, 91, 104, 189, 203–4, 215
  geographical 17
  transport-based 16, 17
social inclusion 17–20, 33, 104, 189, 204
social inequity 187–90, 202–3
social integration 15, 121–3, 131, 135, 137–8, 140, 152, 203–4, 214
social sciences 7, 10–11, 29, 30, 76–7, 99, 100–101, 116, 146, 170, 188
social stratification 7, 8–10, 14–15, 17, 21, 34
South Africa 172, 208–11, 213–15
  Johannesburg 207–9, 211, 216

Soccer World Cup (2010) 1, 207–8, 211, 217
South America 172
Spain 55, 156, 158
spatialisation 15, 21, 101–2
spatialities 29–32, 35
status 2, 3, 8–10, 13, 18–19, 32, 35, 39, 60–61, 82, 111–12, 121, 151, 157, 166, 170–71, 174, 176–7, 179–81, 188–9, 203, 213
  ethnic 188
  occupational 60, 166, 170–71, 174, 176–7, 179
  social status 32, 35, 121, 151, 157, 170–71, 176, 188–9, 203
  work 112
stratification 9, 27, 32–6, 80, 88, 116
students 10, 92, 113
suburbanisation 189, 193, 197–8
Sweden 55, 63, 66
Switzerland 14, 17–18, 100, 105, 122–3, 125, 149–50, 154, 156–8
  Basle 122, 149–50, 156–7
  Berne 122, 148–9, 152
  Geneva 122
  Lausanne 122, 149
  Zurich 2, 99–100, 105–7, 116, 122, 148–9, 154

terrorim 10
The Netherlands 55, 56
tourism 2, 10, 37, 80–83, 91–2, 147, 149
traffic 18–19, 40, 44, 122, 147, 188–9, 204, 207, 210–11, 213, 215
transnationalisation 101, 165–81
  unequal 166–7, 170, 174, 176
transport/transportation 1–3, 7–8, 11, 12, 14, 16–21, 27–33, 35–7, 40–42, 44–5, 76–7, 86–7, 90, 99–101, 103–7, 109, 112, 116, 121, 123, 138–9, 146, 149, 153–4, 159, 169, 177, 187–9, 191, 193–5, 198, 200–205, 207–17
  air 14, 18, 104, 87, 116, 123
  bicycle 14, 116
  car 1, 14, 17–19, 31–3, 40, 42–3, 87, 101, 104, 114, 121, 123, 139, 146, 152, 169, 187, 189, 193–4, 201, 204, 207–10, 213–16

motorised 44, 187–9, 204, 207, 210
non-motorised 207, 216
public 1, 14, 17, 19, 20, 32–3, 103–4,
    106, 107, 112, 116, 121, 154, 169,
    187, 189, 194, 198, 200–201, 204,
    207–8, 210–17
  bus 17, 19, 20, 87, 208–10, 215
  mini-bus 208–9
  taxis 87, 208–13
  train 14, 33, 87, 104, 123, 154,
    209–10, 215
  tram 19, 87, 208
transport planning 19, 40, 99, 102
transport studies 2, 8, 15–18, 20, 28, 30,
    31, 36, 40, 100, 116
travel 11, 14, 17–21, 31–3, 36, 38,
    41–2, 75, 81–3, 86–90, 92, 99, 101–2,
    104–5, 108, 111, 114, 116–17, 121–2,
    127, 139, 155, 190, 193–4, 202, 207,
    209–10, 212–13, 215–16
  communicative 11, 75, 89–90
  imaginative 11
  leisure 31, 102, 108
  physical 11, 75, 90, 104
  social activity 102
  virtual travel 11, 75, 89
travel behaviour 17, 21, 101–2, 116, 190,
    194, 202

Turner, M. 76, 92, 177

unemployment 14, 17, 56, 60, 62–3, 66,
    203
United Kingdom 17–18, 20, 31, 41, 55–6,
    65, 189, 204
United Nations 172
United States of America 54–6, 58, 60,
    65, 158, 187, 189–90
urban development 187–8, 203, 216
Urry, J. 1–3, 7, 10–11, 13–15, 21, 27–9,
    31, 34–8, 43–4, 7–6, 78–9, 85–93,
    99–101, 116, 121, 139, 147, 168
  'meetingness' 76, 85, 88–92
  network capital *see* capital
Uteng, T.P. 27, 39

Vasconcellos, E.A. 207, 210–11, 215

Wajcman, J. 27, 40, 41
Watts, S. 147, 167
wealth 2, 8, 10, 13, 45, 159, 169
Weiss, A. 28, 33, 34
Wellman, B. 102, 121, 140, 169
Widmer, E.D. 121, 123, 136, 147

youth 18, 56–8, 66–7, 114, 194, 197

For Product Safety Concerns and Information please contact our EU
representative GPSR@taylorandfrancis.com
Taylor & Francis Verlag GmbH, Kaufingerstraße 24, 80331 München, Germany

www.ingramcontent.com/pod-product-compliance
Lightning Source LLC
Chambersburg PA
CBHW060831170526
45158CB00001B/137